Cell Culture - Advanced Technology and Applications in Medical and Life Sciences

Edited by Xianquan Zhan

Published in London, United Kingdom

IntechOpen

Supporting open minds since 2005

Cell Culture - Advanced Technology and Applications in Medical and Life Sciences
http://dx.doi.org/10.5772/intechopen.94823
Edited by Xianquan Zhan

Contributors
Soumya Basu, Jyotirmoi Aich, Sangeeta Ballav, Ankita Jaywant Deshmukh, Shafina Siddiqui, Shridhar C. Ghagane, Sridevi I. Puranik, Kalyani Killekar, Rajendra B. Nerli, Murigendra B. Hiremath, Aimen Akbar, Manash K. Paul, Harikrishnareddy Rachamalla, Anubhab Mukherjee, Md. Salauddin, Chenghai Li, Xianquan Zhan, Na Li

First published in London, United Kingdom, 2022 by IntechOpen
IntechOpen is the global imprint of INTECHOPEN LIMITED, registered in England and Wales, registration number: 11086078, 5 Princes Gate Court, London, SW7 2QJ, United Kingdom
Printed in Croatia

British Library Cataloguing-in-Publication Data
A catalogue record for this book is available from the British Library

Additional hard and PDF copies can be obtained from orders@intechopen.com

Cell Culture - Advanced Technology and Applications in Medical and Life Sciences
Edited by Xianquan Zhan
p. cm.

This title is part of the Biochemistry Book Series, Volume 30
Topic: Cell and Molecular Biology
Series Editor: Miroslav Blumenberg
Topic Editor: Rosa María Martínez-Espinosa

Print ISBN 978-1-83969-445-5
Online ISBN 978-1-83969-446-2
eBook (PDF) ISBN 978-1-83969-447-9
ISSN 2632-0983

For EU product safety concerns:
IN TECH d.o.o., Prolaz Marije Krucifikse Kozulić 3, 51000 Rijeka, Croatia, info@intechopen.com or visit our website at intechopen.com.

IntechOpen Book Series
Biochemistry
Volume 30

Aims and Scope of the Series

Biochemistry, the study of chemical transformations occurring within living organisms, impacts all of the life sciences, from molecular crystallography and genetics, to ecology, medicine and population biology. Biochemistry studies macromolecules - proteins, nucleic acids, carbohydrates and lipids –their building blocks, structures, functions and interactions. Much of biochemistry is devoted to enzymes, proteins that catalyze chemical reactions, enzyme structures, mechanisms of action and their roles within cells. Biochemistry also studies small signaling molecules, coenzymes, inhibitors, vitamins and hormones, which play roles in the life process. Biochemical experimentation, besides coopting the methods of classical chemistry, e.g., chromatography, adopted new techniques, e.g., X-ray diffraction, electron microscopy, NMR, radioisotopes, and developed sophisticated microbial genetic tools, e.g., auxotroph mutants and their revertants, fermentation, etc. More recently, biochemistry embraced the 'big data' omics systems.
Initial biochemical studies have been exclusively analytic: dissecting, purifying and examining individual components of a biological system; in exemplary words of Efraim Racker, (1913 –1991) "Don't waste clean thinking on dirty enzymes." Today, however, biochemistry is becoming more agglomerative and comprehensive, setting out to integrate and describe fully a particular biological system. The 'big data' metabolomics can define the complement of small molecules, e.g., in a soil or biofilm sample; proteomics can distinguish all the proteins comprising e.g., serum; metagenomics can identify all the genes in a complex environment e.g., the bovine rumen. This Biochemistry Series will address both the current research on biomolecules, and the emerging trends with great promise.

Meet the Series Editor

Miroslav Blumenberg, Ph.D., was born in Subotica and received his BSc in Belgrade, Yugoslavia. He completed his Ph.D. at MIT in Organic Chemistry; he followed up his Ph.D. with two postdoctoral study periods at Stanford University. Since 1983, he has been a faculty member of the RO Perelman Department of Dermatology, NYU School of Medicine, where he is codirector of a training grant in cutaneous biology. Dr. Blumenberg's research is focused on the epidermis, expression of keratin genes, transcription profiling, keratinocyte differentiation, inflammatory diseases and cancers, and most recently the effects of the microbiome on the skin. He has published more than 100 peer-reviewed research articles and graduated numerous Ph.D. and postdoctoral students.

Meet the Volume Editor

Xianquan Zhan received his MD and Ph.D. in Preventive Medicine at West China University of Medical Sciences. He received his post-doctoral training in oncology and cancer proteomics at the Central South University, China, and the University of Tennessee Health Science Center (UTHSC), USA. He worked at UTHSC and the Cleveland Clinic in 2001–2012 and achieved the rank of associate professor at UTHSC. Currently, he is a full professor at Central South University and Shandong First Medical University, and an advisor to MS/PhD students and postdoctoral fellows. He is also a fellow of the Royal Society of Medicine and European Association for Predictive Preventive Personalized Medicine (EPMA), a national representative of EPMA, and a member of the American Society of Clinical Oncology (ASCO) and the American Association for the Advancement of Sciences (AAAS). He is also the editor in chief of *International Journal of Chronic Diseases & Therapy*, an associate editor of *EPMA Journal*, *Frontiers in Endocrinology*, and *BMC Medical Genomics*, and a guest editor of *Mass Spectrometry Reviews*, *Frontiers in Endocrinology*, *EPMA Journal*, and *Oxidative Medicine and Cellular Longevity*. He has published more than 148 articles, 28 book chapters, 6 books, and 2 US patents in the field of clinical proteomics and biomarkers.

Contents

Preface

Cell culture is a cell cloning technology for large-scale clone cells by simulating in vivo environment conditions such as asepsis, appropriate temperature, and pH, as well as certain nutritional conditions, to enable cell survival, growth, reproduction, and function. It is an important and common technology in cell biology and bioengineering research used to study cell signal transduction, cell anabolism, cell growth, and proliferation. Cell culture includes human, animal, plant, and microbial cell cultures according to cell origin. It can be classified as a primary culture, secondary culture, or cell co-culture per the experiment requirements. Each type of cell culture has its own characteristics and essential conditions. However, one must realize that cell culture is an in vitro method and thus it cannot replace in vivo animal experiments and human trials. Generally, the results of a cell culture must be verified by in vivo animal experiments and further by human trials.

This book focuses on the advanced technology and applications of cell culture in the research and practice of medical and life sciences. Chapter 1 addresses the overall concept and the process of cell culture and its multiple applications and assessment parameters. Chapter 2 discusses the main features of primary cancer cell cultures, the different methods for their selection and management, and their extensive applications in prostate and bladder cancer preclinical treatment processes. Chapter 3 systematically addresses the concept and applications of two-dimensional (2D) and three-dimensional (3D) cell culture. Chapter 4 describes the isolation, expansion, and functional assays of mesenchymal stem/stromal cells, and long-term culture–associated alterations of cellular properties. Chapter 5 discusses stable isotope labeling with amino acids in cell culture (SILAC), and its application in studying ivermectin-mediated molecular pathway changes in human ovarian cancer cells. Chapter 6 discusses the nanotechnology application of mammalian cell culture and its intellectual property rights-related issues in the industrial and academics aspects.

This book presents new advances in the concepts, methodologies, and applications of cell culture. It is designed to stimulate and encourage researchers who develop and use cell culture in different fields in the service of human health.

Xianquan Zhan, MD, Ph.D., FRSM
Professor of Cancer Proteomics and PPPM,
Medical Science and Technology Innovation Center,
Shandong First Medical University,
Jinan, Shandong, China

Chapter 1

A Brief Concept of Cell Culture: Challenges, Prospects and Applications

Md. Salauddin

Abstract

Cell culture is an *in vitro* technique in which cells, tissues, or organs (animal origin) are artificially grown with the support of an artificial environment that encompasses culture medium, CO_2 level, pH indicator, temperature keeping tissues alive and growing appropriately. Organ culture, Primary explant culture, and Cell culture among them cell culture widely used for the understanding of cell growth, normal functions, identification of growth factors, viral vaccine development, recombinant DNA (rDNA) technology, and immunobiological research. Due to high feasibility, cell culture practices highly demandable in the pharmaceutical industry. As well as animal cell culture used in laboratory research to study the cytotoxicity of new drug metabolic studies, aging, therapeutic proteins, the effects of drugs and toxic compounds on the cells and mutagenesis and carcinogenesis. There are a lot of issues in cell culture, Mycoplasma is one of the major. During cell culture, a single antibiotic often cannot kill the mycoplasma. Besides, culture media, pH indicator, incubation, cryopreservation, thawing, passaging of cells, and trypsinization have a great impact on cell culture. This chapter will help the reader to understand the whole process of cell culture and its applications, which will take them one step forward in their virology and cell culture research along with inspiration. This chapter also aids in the concept of cell count, cell suspension, CCF measurement, MOI (Multiplicity of Infection), and cell infection. Eventually, the reader will get a crystal clear concept of cell culture.

Keywords: Chicken Embryo Fibroblast Culture, BHK-21, Cell maintenance, Culture media, Cell culture, Passaging of cells, Cell harvest, Antibiotics, Bulk antigen, Quality control, Vaccine candidate development

1. Introduction

Cell culture or Chicken embryo fibroblast cell culture is a fundamental laboratory technique that widely used in virology, vaccinology, molecular biology, microbiology as well as in biotechnology field. In this *in vitro* study controlled environment like media, pH, CO_2 level, temperature, humidity, O_2 flow mimics *in vivo* condition for cell to grow. Within last past decades a few species have been considered *in vitro* utilizing cell culture technique exposing essential data regarding their biology. The chicken or avian embryo gives an amazing model for the study embryology, developmental biology and production of pharmaceutical proteins in transgenic chickens moreover take part in virology, immunization advancement as well as vaccinology research. Chicken embryo fibroblast cell culture has great impact on veterinary vaccine and biochemical production research. Besides, chicken stem cells have given a perfect opportunity for producing cell-based transgenic birds and a powerful source of cells for vaccine production for poultry and human viral diseases. Besides, they give knowledge into fundamental science namely drug sensitivity testing, cell tracking and cytotoxicity testing [1]. However, various important virus or viral diseases have never been explored at a cellular level so that cell culture technology is still in an early stage of its potential progress. In this document, it is summarized the basic steps engaged in the establishment of primary cell culture and concept of cell culture technique to serve as a practical guide for current and future researcher to leverage the power of cell cultures. This approach has the potential to produce valuable outcomes and suggestions regarding CEF cell culture, cell metabolism, adaptation to different stresses and challenges. Although most published papers discussed on CEF cell culture, BHK-21 and other cell lines, here focused on basic concept and necessary measurement for primary cell culture that inevitably helpful for novice as well as experienced researchers or laboratory personnel of this field [2]. Physiological and physiochemical condition: Nature and impact of pH level, temperature, concentration of O_2/CO_2, and osmotic pressure (culture media) that can be altered the result of the study as well as their effects on the cell culture [3]. Cell culture technology has spread productively within a century; a variety of culture media has been designed. This chapter goes through the brief concept, challenges and current issues of cell culture or chicken embryo fibroblast cell culture technique. Current article does not substance to be a complete guide but rather helps the researcher to plan and make consonant decision to their research experimental or work.

2. Cell culture

General concept of cell culture is the propagation of cells or fibroblast or living tissues in a defined media that conducive their growth. Shortly, a growth of cell artificially known as cell culture. CEF (Chicken Embryo Fibroblast) cell culture is the culture of fibroblast cells obtained from embryo. Embryonated eggs commonly used in the production of bulk antigens, vaccines and other biochemical. SPF (Specific Pathogen Free) eggs obtained from SPF (Specific Pathogen Free) chicken flocks which have been intensively monitored for infectious agents and have not been vaccinated; or, where justified (e.g. for production of some inactivated vaccines) and in line with the marketing authorization, from healthy chicken flocks. For the propagation of virus laboratory personnel or researcher should have to choose specific route of inoculation of the SPF egg based on the study microorganisms (virus) that is being propagated [4].

Baby Hamster Kidney cells (BHK-21) are generally used in life science research work and the biopharmaceutical industry. Chinese hamster ovary (CHO) cells, baby hamster kidney (BHK) cells, mouse myeloma cells comprising NS0 and SP2/0, hybridomas, and human cell lines (HEK293, HT-1080) are the very commonly used mammalian cell lines at large scale [5]. Among these cell lines, BHK-21 has found applications at large scale in veterinary viral vaccines against foot and mouth disease (FMD) and rabies virus, as well as heterologous protein production (Factor VIII) [6]. The BHK-21 cell line was established in 1961 from the kidneys of 5 Syrian hamsters from litter number 21. Since this time, this cell line has been considered as a research facility standard for the development of countless viruses and the study of numerous biological processes. Baby hamster kidney (BHK) cells are one of many different vertebrate cell types used for the propagation of viruses by infection and transfection [7]. Generally obtained from BHK cells can vary widely if not obtained from the ATCC. For the production of animal products (vaccine) baby hamster kidney (BHK) cells are generally used, the most important of which is the production of a vaccine against FMD and rabies. Also, BHK was used in the production of recombinant proteins, such as blood coagulation factor VIII for the extraction of DNA from Pseudorabies virus and production of capture antigen enzyme-linked immunosorbent assay (ELISA) when diagnosing Japanese Encephalitis (JE) [8–10]. BHK-21 cells were grown and propagated in modified minimum essential media (MEM) by adding a set of proteins including Lactalbumin (2.50 g/l), yeast extract (1.00 g/l), peptone (2.50 g/l), New Zealand casein (1.00 g/l); Glutamine (0.50 g/l) and 2% sodium bicarbonate were also added to the culture medium. Also, 5–10% of the serum treated with polyethylene glycol 6000 was added to modified MEM. Penicillin G and streptomycin was added at the rate of 100 IU to control microbial load. After keeping the BHK cells in a liquid nitrogen tank and passing them through the preparation steps, the cells were transferred to CCF for subsequent cultivation. The flasks were kept in incubators at an operating temperature of 36°C or 36.5°C or 37°C and 5% carbon dioxide. Maintaining the starting pH of the culture ranged from 7.1 to 7.2 or 7.4 while carbon dioxide was used to control the pH. Common morphology of some cells are *Fibroblastic (or fibroblast-like) cells*: Shape-bipolar or sometimes multipolar, elongated, that attached to a substrate during growth and wrench commonly during culture. *Epithelial-like cells*: Shape-polygonal, Dimensions-regular, and which attached to a substrate in discrete patches. *Lymphoblast-like cell*: Shape-spherical, Commonly, grown in suspension and do not attach to the surface of flask.

3. Basic steps of cell culture technique

Cell culture or CEF cell culture is crucial laboratory work. During cell culture laboratory personnel or researchers have to maintain chronological workflow step by step. By maintaining the following steps cell culture should be done very smoothly. For CEF cell culture, SPF eggs were collected from authentic sources with a legal document. Then incubate for the recommended date. Afterward-

1. Personal hygiene and safety management

2. Sterilization and disinfection

3. Arrangement of instrument and appliances as well as power supply

4. Biosafety cabinet (BSC) management

5. Management of essential ingredients and chemicals like Media, FBS, Trypsin, Phosphate Buffer Saline (PBS), Tincture of iodine, Antibiotics, etc.

6. Management of Pipettes, Burette, Tips, Waste disposal box, Aspirator, Tissue

7. Chopping, washing embryo, and filtering (for CEF cell culture)

8. Centrifugation

9. Cell suspension and stock management

10. Cell count and splitting of the cell

11. CCF, Media, and incubator management

12. Cell line development

13. Cell observation, harvest

14. Cell count, subculture, infection, and bulk antigen production

4. Fundamental parts of cell culture technology

4.1 Culture medium

Culture medium is a composition of nutrients and selected buffer that helps to grow an organism naturally. Media can be designed based on variety of cell, types of cells because it is necessary for cell survival, proliferations and growth. The influence of cell culture technology creates inevitable progress in molecular biology research. This technique widely utilized in different fields like the assessment of toxicity and efficacy of new drugs, development of various biopharmaceutical products and vaccines, and used in reproductive technology. No one probably would argue against the claim that a culture medium is the foremost essential measure in cell culture technique. Selection of suitable media for research goal is essential. Sometimes researchers should modify a properties or composition of medium in order to their experiment. There are mainly two types of media used by researcher such as natural media and synthetic media [3, 11, 12]. *Natural media*: The media that comprising of natural biological substances, e.g. embryo extract, plasma, and serum. Coagulant or clots i.e. plasma which is separated from heparinized blood, serum, and fibrinogen. Tissue extracts and Extracts of chicken embryos, liver, and spleen and bone marrow extract considered as a natural cell culture media. Biological fluids such as Plasma, serum, lymph, amniotic fluid, and pleural fluid also used as culture media. *Synthetic media*: The media comprising with basal medium and supplements, like-serum, growth factors, and different hormones. On the basis of type of supplements synthetic media can be classified into a few bunches such as: (1) Serum-containing media: In this media as a supplement bovine, human, equine, or other serum is used. This media causes less reproducibility and microbial contamination due to unknown and unclear media composition. (2) Serum-free media: This media has a high reproducibility due to presence of the supplements of bovine serum albumin or α-or β-globulin (crude protein fractions). (3) Xeno-free Media: Prepared based on human-source components i.e. human serum albumin, are commonly used as supplements but there are no animal components used as supplements. (4) Protein-free media: Prepared based on undefined components like- peptide fractions (protein hydrolysates) which are generally used as supplements, and (5) Chemically defined media: Undefined components are used as supplement. Crude protein fractions, tissue extracts, hydrolysates are not good supplements, but highly purified components (recombinant proteins) are appropriate supplements. Followings are some brand name of culture media commonly used in the laboratory: Eagle's Minimum Essential Medium (EMEM) a simpler basal medium (BME) for culturing mouse L cells and HeLa cells. Dulbecco's Modified Eagle's Medium (DMEM) is a basal media (BME). Iscove's Modified Dulbecco's Medium (IMDM) used for growth of lymphocytes and hybridomas. Hybri-Care Medium is a combination and modification of DMEM considered for the propagation of hybridomas and other fastidious cell lines. Ham's Nutrient Mixtures used for Chinese hamster ovary (CHO) cells. Kaighn's modification of Ham's F-12 (Ham's F-12 K) was designed to develop and differentiation of primary cells with or without serum. DMEM/F12 Medium is a 1:1 mixture of Dulbecco's modified EMEM and Ham's F-12 for the growth of a wide range of cell types in both serum and serum-free formulations. Leibovitz's L-15 Medium for biopsy samples without CO_2.

4.2 pH value

pH plays a vital role in cell culture. The cell growth rate is decline associated with Fluctuations in pH level. That's why routine monitoring is essential. For the cells is 7 the optimal pH, and decline or increase in pH can hinder the growth of cells. More decrease in pH level (usually in between 6.0–6.5), can stop the growth rate of the cells, and cells are start losing viability at low pH level. That's why pH level should be maintained and monitored carefully for individual cell line. If pH level fall rate is less than 0.1 units/day, that indicates the cell condition is good and no need to hurry to change the culture medium immediately. If it is 0.4 units/day (pH drop rate), that indicates the culture medium need to be changed quickly [13]. Alkali (like- NaOH, KOH) or acid (HCl) solution helps to control pH level in culture medium. Besides, $NaHCO_3$ (sodium bicarbonate) or natural buffer solution, and addition (need base) of CO_2 gas to the bioreactor also helps to maintain optimum pH level in culture medium. Generally, pH electrode (silver chloride electrochemical-type) used within the bioreactor [14]. For the proliferation of cell in culture medium an optimum, stable as well as balanced pH is essential. Depending upon cell type and culturing process pH level may vary and specific. Generally, 5–10% CO_2 required to grown cells using buffered media that contain $NaHCO_3$ (sodium bicarbonate) and where maintained the range of pH 7.2 to 7.4 [15]. CO_2 incubator, optimum pH level, ideal temperature, optimum moisture condition, sterile and clean working environment are essential to maintain and complete an experiment. In the cell culture medium, the carbonate buffer helps to hold constant pH and take parts in releasing CO_2 gas in the CO_2 incubator. Color of the culture medium depending and changing with the pH level of culture medium. Color indicates altering the medium and CO_2 levels [3]. Commonly, 4–10% of CO_2 is practiced in the cell culture technique. By maintaining HCO_3^-concentration and CO_2 tension in culture medium one can easily achieved optimum pH and osmolality [16]. By observing the color of the media can easily identify the pH condition like- Phenol red to yellow/ orange color indicates too acidic where pH 6.8 (bellow), Red to pink color indicates pH above 7.0 to 7.7 which is normal, and bright fuchsia color indicates pH 8.0 to 8.2 (too alkaline).

4.3 Effect of temperature

Generally, cell culture need 37°C for incubation called control temperature. Proliferation and multiplication of cells are significantly decreased at more than 40°C, like 41°C or 42°C temperature, and increased temperature may also cause high apoptotic rate of CEF cells. Cell viability, apoptosis, proliferation, and oxidative status of cells in culture medium can be altered with high temperature. ROS (reactive oxygen species) formation increased with increasing Temperature (Proportionally) [17]. During the transportation must be care full about temperature. Besides regular or routine inspection is recommended for better results. In cell culture technique one of the most challenging issues is to grow cell, that's why an ideal temperature play a vital role in cell culture along with good supplementation of nutrients. For the cell division the optimum temperature should be vary on cell type that assist to maintain growth rate. Generally, at optimal temperature metabolic function of cell is optimum as well as good that helps to increase their size, and proliferation rate [18]. Temperature requirement varies based on cell type like- *S. frugiperda* (sf-21, *Trichophusia ni* (Tn-5), *S. frugiperda* (sf-9) (insect) need 27°C, Chinese Hamster Ovary (CHO) (Mammalian)-need 36–37°C, Human embryonic kidney (HEK- 293 T) (Human) need 37°C, HF 205 and HF 210 (plant) need 27°C.

4.4 Incubation

A device in which microbial culture or cell culture is grown and maintained with customized temperature, humidity (relative humidity 95%), oxygen, CO_2 level, and other conditions. For virus cultivation, cell culture as well as cell infection, vaccine development incubation is a very much crucial and fundamental issue. The incubator is an essential instrument in cell biology research, Microbiology, Biotechnology, Molecular biology research. On the other hand, an egg incubator is one of the most important for embryonated egg production in the laboratory and large scale. BOD (Biological Oxygen Demand) incubator is popular in this regard. There are so many necessary points always bear in mind to maintain good incubation. Firstly, cleaning inside, cabinets, outside and handle that helpful to eradicate cross-contamination, also helps to hold a good quality of cell, media, SPF (Specific Pathogen Free) eggs, and other chemicals or ingredients. Logbook maintenance is another vital issue to maintain high-quality research work. Through logbooks, laboratory personnel can easily identify any issue related to the incubator. A dedicated power supply essential to maintain cell quality, growth, metabolism as well as important to maintain cell physiology. CO_2 level and the supply are very much essential to maintain moisture and pH (normal range is 7.2 to 7.4). Generally, 5% CO_2 with 37°C is used to maintain the cell. Eventually, inventory management will helpful for GMP (Good Manufacturing Practice) and proper cell culture as well as laboratory work.

4.5 Cryopreservation

Cryopreservation is the method in which intact living cells are conserved as intact in liquid nitrogen at cryogenic temperatures. On the other hand, in cryopreservation system using low temperature helps to protect living cells structurally intact. Freezing system keeps the living cells frequently for a long time (often for years). Freezing temperature ceased their typical metabolic activity that's why cells are protected from damage caused by long time preservation, and chemical reactivity. Gently handle cells because cell may be damaged and will get stress during the freeze–thaw process. Optimal cryopreservation of cells relies on proper freezing and thawing methods. A successful cryopreservation method calculating based on recovery rate of cells (frozen) from low temperature, percentage of alive cells, and rate of cells that function as normal after thawed [19]. Basically, for cryopreservation harvest the cells in exponential growth. Then gentle centrifugation done at 125 × g for 10 minutes. After that check and scree the media. The procedure starts with taking freezing medium (GM-which warm at room temperature for 30 minutes) that containing a cryoprotectant such as Dimethyl sulphoxide (DMSO) (e.g. 5–10% v/v DMSO), fetal bovine serum (FBS) (10–20% v/v) and at high cell density ($1–5 \times 10^6$/ml) and sometimes added knock-out serum replacer (KoSR; 20% v/v final), bovine serum albumin (5% final) or human serum albumin. From some recent research [20–22] it is said that freezing rate has great impact on viability of cryopreserved cells. It is suggested that cells be slowly cooled (like 2°C, −20°C, −80°C for at least 24 hours and finally preserved in liquid nitrogen at −196°C) that gives better surviving rate of cell in the cryopreservation process. Record logs must maintain during all the steps. From the final cryovial (that contain cell at −196°C) remove one vial and restore the cells in culture medium to determine cell viability and sterility. Recovery rate of cryopreserved cells depends on the types of cultured cell. Some cell needs several days, some shows low viability on the day of culture, in some cases cell produce debris, some cells are shows normal viability after 24 hours' post-thaw. Before retrieving of cryopreserved cells clean the biosafety cabinet, prepare the CCF, media, FBS and arrange all the instruments and appliances (sterile). Then wash the cryovial with 70% ethanol and place it in a water bath for 2 to 5 min at 37°C to melt. Transfer the thawed cell in a tube and gentle centrifugation (10 minutes at 125 × g) needs to discard (supernatant) cryoprotectant in the meanwhile collect the cell pellet and suspend the cells in 1 mL or 2 mL of complete growth medium (GM) then proceed for cell count and subculture in new CCF for 24 hours' observation. In the process of cryopreservation, significant rate of cell survival and maintenance of cell integrity (structural and morphological) can be achieved by using cryoprotective agents (CPAs). Excipient is an ingredient added intentionally to the drug substance which should not have pharmacological properties in the quantity used. Commercially available CPAs namely Dimethyl sulphoxide (DMSO) is commonly used as CPA. Factors behind the success of cell survival [23] are: (a) Type and concentration of cryoprotectants (an additive, such as glycerol or dimethyl sulphoxide, that can protect cells against freezing injury). (b) Cell density in cryopreservation solution at the time of freezing. (c) Cooling and thawing rates of cell suspension. (d) Dilution rate of thawed cell suspension. The main advantages [24] of cryopreservation are easily found original cell lines from the safety stocks, preserve the cells for year after year and lastly smoothly perform continuous research or experiments.

4.6 Thawing

Correctly thawing of cells is crucial to recover quickly, yielding the highest viability and functionality possible. Some cryoprotectants (e.g. DMSO) has toxic effect on cells, due to the possibility of toxicity cells should be thawed rapidly and not allowed to remain in the freezing medium no longer than required time. Firstly, retrieve the cryovial containing the frozen cells from liquid nitrogen (−80°C or – 196°C freezer) and immediately place it into a 37°C water bath (for 1 to 2 min) or place immediately in a pre-equilibrated thermo-conductive rack or tube module resting on dry ice to minimize cell warming/thawing. Rapidly thaw the cells (< 1 minute) by carefully swirling the cryovial in the 37°C water bath up to there is a little bit of ice left within the cryovial. Then place the cryovial into a BSC. Gently wipe the outside of the cryovial with 70% IPA (Isopropyl Alcohol) prior to open the cryovial screw. Carefully add (dropwise) required amount of pre-warmed GM into the tube (centrifuge) that containing recently thawed cells. Place the cell suspension in centrifuge machine for centrifugation and set 200 × g for 5–10 minutes (it may vary based on cell type). Inspect the transparency of supernatant and visibility of a pellet at the bottom after completing the centrifugation. Discard the supernatant aseptically without breaking the pellet. Softly resuspend the cells by gently pipetting with GM and prepare required concentration, then transfer the cell suspension into the CCF, and place it in the suggested culture environment. Inspect the cells using an inverted microscope for morphology. Examine an aliquot of cells for the ability to exclude trypan blue. If cells pass both inspections, they are ready for culture [25].

4.7 Passaging of cells

Subculture of cell commonly known as passaging of cells and the ratio of subculture is 1:2. The main concept of passaging: cells are split into half in each subculture. Continuous cell lines should will be passaged with higher split ratio due to their higher replication rate. Usually the number of times the cells have been subcultured into a new CCF known as passage number. In the case of diploid cell cultures, the number cell passage is partially equal to the number of population doubling level (PDL) since the culture was begun. On the other hand, PDL of continuous cell lines is not fixed like diploid cell culture. Mostly the PDL is an estimation or prediction. PDL may ups and down with cell stress and cell death (due to necrosis, apoptosis). Loss of proliferation capacity of cell, contamination of culture medium may also responsible. A common formula for the calculation of population doubling level: PDL = 3.32 (log Xe – log Xb) + S; where Xb is the cell number at the beginning of the incubation time, Xe is the cell number at the end of the incubation time, S is the starting PDL. Another common formula used to calculate the population doubling level is [26]: Log (*N/N*) x 3.33; Where N is the number of cells in the culture vessel at the end of a certain time interval, and N is the original number of cells plated in the vessel, Population doubling and passage number are often mixed up or thought to mean the same thing. The passage number describes the number of times that a culture has been subcultured The population doubling time is important to know when the number will be double and passaging will be required. According to ATCC [24] the population doubling time calculate with the formula: DT = T ln2/ln(Xe/Xb); where T is the incubation time in any units, Xb is the cell number at the beginning of the incubation time, Xe is the cell number at the end of the incubation time.

4.8 Trypsinization

Trypsin is an enzyme that is used to remove adherence proteins from a cell surface. Generally, trypsin-based disaggregation so-called trypsinization. Disaggregation of cells from the CCF commonly crude trypsin used, the effect of raw (crude) trypsin can easily be neutralized by commercially available serum (FBS) or trypsin inhibitor. On the other hand, pure trypsin is also used in the cell degradation process which is less toxic and very specific in action [27]. Commonly 0.05% trypsin used in laboratory work. Sometimes trypsinization causes cell damage and sometimes may not effective for some cells thus other dissociation agents (enzyme) are recommended for the dissociation of cells. Warm and cold trypsinization are the two common approaches. An extensively used method is warm trypsinization. In which cells are washed with basal salt solution and then add warm trypsin (37°C) adequately and stirred properly. The supernatant dissociated, the cells are dispersing in the medium. In the case of cold trypsinization, cellular damage is reduced, resulting in a high yield of viable cells also improved survival rate. For cold trypsinization cells are maintained in ice after washing with media or salt then treated with cold trypsin for 6–24 h. After that remove and discard the trypsin and incubate the CCF at 37°C (for 20–30 min). Dispersion of cells may start and fully dispersed cells counted using hemocytometer then dispersing in a medium for further use. The easiest way of trypsinization is a). Discard the media from CCF b). Wash the cell surface with PBS (4 ml for 75 cm^2 CCF) c). Take trypsin (Room temperature) d). Rinse monolayer of cell with trypsin–EDTA (2 to 4 ml for 75 cm^2 CCF) e). Stay for 2 minutes then discard trypsin and Incubate the CCF at 37°C for 5 minutes f). Tapping the CCF and collect the cells by scrapper stored in a tube g). Spin down the cells, resuspended by adding the growth medium or fresh medium, and Count the cells h). Split into a new flask and Incubate at 37°C.

4.9 CCF measurement

CCF denotes a Cell Culture Flask. It is also known as a tissue culture flask. CCF is important for culturing of cells, transportation of cells, and media. There are a lot of different volumes of CCF used for research work. Commonly, used flask volume are 25 cm^2, 75 cm^2, 175 cm^2, 225 cm^2, 300 cm^2. Commonly 75 cm^2 CCF preferable for laboratory work. Cell contamination generally appeared during cell culture laboratory. Proper knowledge of CCF handling can minimize cross-contamination, which improves the quality and physiology of the cell. Rough handling of CCF during media transfer, passaging, scrapping will be responsible for the different vital issues. Mycoplasma contamination is one of them. Prompt and Improper pipetting during cell harvest and split from CCF may cause stressful conditions on cells and resulting in cell death. Scrapping of the cell for subculture or infection or cell count gently handles the CCF, corkscrew. Mild flame spark on corkscrew (CCF) by flame gun or gas burner helpful to safe the cell and flask environment. Covering the cork with Paraflim very much essential to save the cell. IPA (Isopropyl Alcohol) spray must be done before and after handling of CCF. After application of IPA, then CCF, media, FBS (Fetal Bovine Serum), tips, flame gun, trypsin other materials, and appliances allow entering into BCS (Biosafety Cabinet) for further processing. Some points should be bear in mind regarding CCF such as, is CCF allow pipettes, tissue scrapper, tissue spatula properly? Is CCF has marked on both sides? Is there any leakage? Is there any crake on cork? Is the bag of CCF tightly pack? Is the CCF clean (inside)? These points might be helpful for cell culture.

4.10 Cell suspension

The cell suspension is nothing but suspension culture. It's another type of cell culture where a small amount/volume of cells is permitted to grow in growth media forming suspension called cell suspension. If the cells are derived from other cultures or homogenized tissue, then use suspension culture. Both suspension culture and adherent culture are the same.

4.11 Cell count and cell infection

Cell counting was performed using a hemocytometer (Neubauer improved counting chamber, Precicolor HBG, Germany) or MacMaster slide and Trypan blue exclusion every 24 h (1,1 mixture of 0.2% Trypan blue in normal saline solution and sample). After placing the stock cell suspension on the hemocytometer and place a coverslip on it. Count cell of 4 (16 × 4) site, then the average of 4 sites of hemocytometer and count the total cell as = Average (number of the cell) × 10,000 × 2. Cell culture flask such as 25 cm^2 contain 5 to 10 ml culture media, 75 cm^2 contain 10 to 30 ml culture media, and 175 cm^2 contain 40 to 150 ml culture media. One (1) cm^2 need 90,000 cells likewise 75 cm^2 need 75 × 90000/5 = 13,50,000 cells minimum. Viable cells are considered as unstained ones and while stained cells are considered as dead under the inverted microscope. Cell counts are important for monitoring cell health and proliferation rate, assessing immortalization or transformation, seeding cells for subsequent experiments, transfection or infection, and preparing for cell based assays. Cell counts must be accurate, consistent, and fast, particularly for quantitative measurements of cellular responses. Cell infection is required for virus propagation, bulk antigen production. MOI rate is very much essential in cell infection. There are three types of MOI commonly used in the laboratory such as 1 MOI, 0.1 MOI, 0.001 MOI where 1 MOI means one (1) virus can infect one (1) cell. 0.1 MOI denotes 10 cells infected by one (1) virus and one virus can infect 100 cells in 0.001 MOI. Generally, practice 0.1 MOI means one virus is enough to infect 10 cells.

4.12 PFU (PFU), ELD_{50}, EID_{50}, MOI, $CCID_{50}$, $TCID_{50}$

Plaque forming units (pfu) is an assessing of the total number of infectious virus particles. It is ascertained by a plaque-forming assay. In the field of virology study, a plaque-forming unit (PFU) is a measurement of the number of particles capable of forming plaques per unit volume i.e. virus particles. It is a functional measurement rather than a measurement of the absolute quantity of particles: viral particles that are defective or which fail to infect their target cell will not produce a plaque and thus will not be counted. For instance, a solution of virus with a concentration of 1,000 PFU/μl indicates that 1 μl of the solution contains enough virus particles to produce 1000 infectious plaques in a monolayer cell, but no inference can be made about the relationship of pfu to the number of virus particles.

ELD_{50}- Embryo Lethal Dosage. ELD_{50} unit is the amount of virus that will kill 50 percent of inoculated eggs.

EID_{50}- Embryo Infective Dosage. EID_{50} unit is the amount of virus that will infect 50 percent of inoculated eggs.

Multiplicity of infection (MOI) is the average number of virus particles infecting each cell. MOI is related to pfu by the following formula: Multiplicity of infection (moi) = Plaque forming units (pfu) of virus used for infection/number of cells. For example, if $2x10^6$ cells is infected by 50 ml of the virus with a titer of 10^8 pfu/ml. The moi will be $0.05*10^8/2*10^6 = 2.5$. The fraction of cells that are not infected is $P(0) = 1 - e^{-moi}$. To ensure 99% of cells are infected requires moi > 4.6. Assume the conditions used for plaque assay and TCID assay do not alter the expression of infectious virus. $TCID_{50}$/ml and pfu/ml are related by pfu/ml = 0.7 * $TCID_{50}$. As a working estimate, one can use pfu/ml = 0.5 * $TCID_{50}$ [28].

$CCID_{50}$: Cell culture infectious dose which will infect 50% of the cell.

$TCID_{50}$ is the tissue culture infectious dose that will infect 50% if the cell monolayers are challenged with the defined inoculum. Two methods commonly used to calculate $TCID_{50}$ (can also be used to calculate other types of 50% endpoint such EC_{50}, IC_{50}, and LD_{50}) are a). Spearman-Karber [29] b). Reed-Muench method.

4.13 Plaque assay

Plaque-based assays are the standard method used to determine virus concentration in terms of infectious dose. Viral plaque assays determine the number of plaque-forming units (pfu) in a virus sample, which is one measure of virus quantity. This assay is based on a microbiological method conducted in Petri dishes or multi-well plates like 6 well or 24 well etc. Specifically, a confluent monolayer of host cells is infected with the virus at varying dilutions and covered with a semi-solid medium, such as agar or carboxymethyl cellulose, to prevent the virus infection from spreading indiscriminately. A viral plaque is formed when a virus infects a cell within the fixed cell monolayer [30]. Virus quantification involves counting the number of viruses in a specific volume to determine the virus concentration. In research and development (R&D) based commercial and academic laboratories, the production of viral vaccines, recombinant proteins using viral vectors, viral antigens and clone screening, multiplicity of infection (MOI) optimization, and adaptation of methods to cell culture all require virus quantification. For quantification of virus incubated after infection at 37°C in a 5% CO_2 incubator at 6, 12, 24, 36, 48, 60, and 72 h post-inoculation (hpi) based on the requirement to visualize plaques in wells [31, 32]. To quantify virus there are a lot of other methods used such as Focus forming assay (FFA), Endpoint dilution assay, Protein assays, Hemagglutination assay (HA), Bicinchoninic acid assay, Single radial immunodiffusion assay, Transmission electron microscopy (TEM).

4.14 Cell harvest

Knowledge of splitting, media, trypsinization, cell handling is essential to harvest cells. Firstly, remove and discard the media from CCF. For 75 cm^2 CCF, mild PBS wash is required before the application of warm trypsin or trypsinization process. Add trypsin (2–4 ml/75 cm^2) then incubate as like trypsinization process. When detached cells appear then add 2–5 ml growth media (GM) to inactivate trypsin. Gently pipette to disperse the medium to ensure recovery of >95% of cells. Sometimes commercial trypsin inhibitor is added. Carefully centrifuge the collected cell suspension at 300–1000 X g for 5–10 min. Discard the supernatant and add GM at the required amount for the preparation of cell count. Split the cells after counting or go for further processes that need. There is a lot of problems that may be appeared like detachment difficulty of cells from culture flasks, cell adherence difficulty, insufficient attachment of cells, low viability of cells, clumping after detachment, damage of cell membrane, and cell death. The possible solutions to the above problems will be careful during the following such as a). Check the quality, date, and concentration of trypsin before use b). Be careful during antibiotic application if any c). Splitting, media replacement required before harvesting if cells are in stress d). Avoid vigorous pipetting and long centrifugation.

4.15 Antibiotics

Routine cell and tissue culture according to good cell culture practice (GCCP) [33] should not require the use of antibiotics as they can never be relied on as a substitute for effective aseptic techniques. However, its use is still widespread e.g., OECD TG 432 [34] due to established routine procedures in many laboratories. Antibiotics are agents that may arrest or disrupt fundamental aspects of cell biology, and, while they are effective against prokaryotic cells (i.e. Bacteria), they are also capable of causing toxic effects in animal cells. Not surprisingly, antifungal agents, being directed at higher order, eukaryotic microorganisms, are likely to be more toxic to animal cell cultures. In addition, antibiotics often make it more difficult to detect microbial contamination. These obvious contraindications, the use of antibiotics in cell and tissue culture should be focused in two areas: a) Protection of materials at high risk of contamination such as tissues, organs, and primary cultures in cases where sterility cannot be guaranteed, and b) The positive selection of recombinant cell clones based on the expression of antibiotic resistance genes [33]. If antibiotics are needed, a justification for the use of antibiotics in the procedure is suggested.

4.16 Laboratory management

Good laboratory setup is essential for cell culture as well as good laboratory practice is also essential for better and smooth work. The following discussion and points are very much important for a cell culture laboratory and are also supported by Maneesha et al. [35] and Coecke et al. [36].

Location and ideal layout with the purpose-built facility is the first priority. Room data sheet (RDS) in which available all the data of every facility, specification of the room define its location, a number of doors, windows, pass box, ventilator, light, air conditioner, fire alarm like everything present on that room permanently. Specification of instruments must be available at the working area that helps the laboratory personnel to operate it. Standard operating procedure (SOP) helps to do research/ laboratory work in a defined way to get a better outcome. Define the area of a room with a specific class and biosafety level by the standard of ISO, GMP. Define or specify works of laboratory personnel with defined working areas according to CDC, NIH-USA also beneficial for a good outcome. There are four biosafety levels based on hazard or pathogenicity or virulence of microorganism and toxicity of agents. The basic biosafety level known as biosafety level 1 (BSL-1). In this level generally very common research work done with easy protection. Normally in BSL-1 working with those organisms or agents which are not harmful to healthy person. In biosafety level 2 (BSL-2) working with those organisms or agents which is known as moderate-risk agents, and known as a potential threat to human resulting produce disease (by ingestion or through percutaneous or mucous membrane exposure) of varying severity. Generally, cell culture should be performed at BSL-2 laboratory. Sometimes the biosafety level depends on the type of cell line and working style. In biosafety level-3 (BSL-3) working with that agents which have the capability to transmit through air (aerosol transmission). BSL-3 agents may be indigenous or exotic, having potential threat to human, may cause serious health issue, and may be potentially lethal. Biosafety level-4 (BSL-4) deals with exotic agents that create life-threatening disease of an individual through infectious aerosols and for which no treatment is available. These organisms or exotic agents are restricted to high containment laboratories. The easily accessible facility of the emergency shower should be available in the laboratory area. Access control in laboratories should be helpful to maintain unwanted occurrences. A separate logbook is very much helpful for liquid nitrogen and CO_2 management. Ventilation and pressure control like negative and positive pressure controlled area must be defined, HEPA filtered providing positive pressure to clean areas, is recommended where space and resources allow. Electricity room (uninterrupted power supply (UPS) units should be provided for essential equipment (class II cabinets, incubators, air filtration) and to allow cell culture procedures to be completed. Accountability for all the staff will provide a smooth working environment. Documentation, Training (Fumigation, 5 s, SOP, etc.) and Monitoring of staffs, Emergency service provider contract with the third party are essential for good laboratory management. Before receive and entering all the reagents, chemicals, and supplier's documents must be checked by maintaining a logbook. For the management of inventory and documents should be established hard copy or electronic form that stored the information of materials, cells, suppliers, overall all the possible information. Staff safety is a vital issue. The primary concerns regarding safe management of liquid nitrogen storage are frostbite burns from skin contact with liquid nitrogen and asphyxiation due to exposure to low oxygen levels when nitrogen gas is released from vessels. Finally, all service personnel entering the laboratory should receive instruction in special laboratory hazards and any necessary procedures for working in clean areas e.g. gowning, hand disinfection. All the person needs to use a separate biohazard bag/bin to ensure safe hazard management. Like- Infectious non-sharp waste (incineration/deep burial)-Yellow bag; Plastics and sharps (chemical treatment/autoclaving/shredding/microwaving)- Blue bag; Infectious non-sharp waste (chemical treatment/autoclaving/microwaving)- Red bag; Incineration ash and solid chemical wastes (secure landfill)- Black bag.

4.17 Challenges during passage of cell

During cell culture, laboratory personnel and researcher should follow up the cell routinely. Splitting of the cell depends on cell doubling number, cell type, pH level, media, and so many cell culture-related issues. There is a lot of challenge situation faced by laboratory workers. The most common problems that cause major issues in the laboratory may also ruin the running works are misidentification of cell line, Contamination of culture and media, Rough handling, Poor cell growth, Poor cell attachment, Improper trypsinization during harvesting, Improper cell count, Improper split ratio during the passage, Clumping of cell, Cell death. Incubation time, temperature, and CO_2 level also have a great impact on cell culture as well as on subculture.

4.18 Maintenance of cell

The routine follow-up of cell morphology is necessary. To maintain a good cell line routinely change of the medium is essential for the both proliferating or non-proliferating cells. The culture medium should be changed repeatedly in the case of proliferating cells compared to the non-proliferating one. The rate of cell growth, cell morphology and metabolism of cell indicates the urgency and time interval of medium change. For example, HeLa cells are rapidly growing transformed cells, in the case of HeLa cell the culture medium should be changed twice within 7 days, whereas for slowly growing non-transformed cells (like IMR-90 cells) the culture medium may be changed once in a week. Continuous cell lines, Chicken embryo fibroblast cell (CEFC), Embryonic cells and transformed cells develop quickly that's why these cells need rapid sub-culture and altering the culture medium. While normal cells are grow slowly. Generally altering the medium depends on pH level. Immediately change whole medium when the pH level appeared 7.0, cells are stop proliferating at pH 6.5, and the cells may lose their durability and viability when the pH level drop in between 6.5 to 6.0. The drop rate of pH is commonly estimated for each and individual cell line with a selected culture medium. If the drop rate of pH is less than 0.1 units/day, that indicates no harm and no need to hurry to change the culture medium immediately. When the drop rate of pH is 0.4 units/day, that indicates the culture medium need to be changed immediately [13]. A laboratory person or a researcher can easily maintain cells by maintaining SOP of cell handling and culture procedure, by counting passaging time because 10–30% density of cell is standard but at 80–90% density cell should be split as well as cell count may be helpful in this regard. Must pay attention to media quality, color, clarity, foul smell results from infection of the cell. Cell health and cell concentration, appearance observed regular interval by bright field microscope with 20–60x magnification may help to maintain cells and eradicate clumping, detachment, apoptosis. Quality control (QC) documents, Certificate of Analysis (COA), Cell transportation SOP are crucial to maintaining cells. Logbook entry for all the daily activities like pH level daily basis, media condition, temperature, CO_2 level, assigned peoples information, cell condition, passage number, all the information about the cell very much essential to maintain cell for either small or large scale work.

4.19 Mycoplasma: a issue

Mycoplasma contamination is a serious and widespread problem in cell culture. Mycoplasma is often passed from culture to culture and from lab to lab. Mycoplasma can ruin whole research if data collected from mycoplasma-infected cells or cultures. Among all the contaminants (biological) in the laboratory mycoplasma have the capability to spread rapidly and causes detrimental effect on cells because of their detection rate is very low as well as their serious impact on cell lines. In spite of the fact that mycoplasmas are actually microscopic organisms (like bacteria) but they have some particular characteristics that make them identical. Mycoplasmas can easily survive and multiply at high densities without producing any noticeable signs. They are very harmful to any cell culture. Mycoplasmas can easily alter the host cells' metabolism and morphology, cause chromosomal aberrations and damage of cell that provoke cytopathic effects. Mycoplasma ought to be tested at least once a month is recommended in laboratory and research work. Two different testing methods, such as DAPI staining and PCR are helpful but a commercially available mycoplasma kit is also recommended for the detection [37]. A routine screening process might be helpful to eradicate mycoplasma contamination from the lab. The following points are essential to prevent mycoplasma issues [38]. a). Wearing personal protective equipment (PPE) during cell culture that includes a dedicated, clean lab coat and gloves b). Checking the COA, QC pass of cells' origins c). Ensure proper sterilization d). Always clean the working area e). To avoid cross-contamination work with only one cell at a time f). Always ensure covering the media bottles and CCF. Do not use the hood for storage and work always within biosafety cabinet g). Should be cautious in the use of antibiotics because it is reported that antibiotics have no impact on mycoplasmas h). Maintain logs for record-keeping that help to identify possible contamination sources if needed i). SOP develop for routine mycoplasma screening.

4.20 Bulk antigen and vaccine candidate development

A large scale of virus production is known as bulk antigen production. A Proper culture of the cell, passaging, infection of cell-based on required MOI by a specific virus, harvesting, filtration of the virus, QC test, and COA gives the final confirmation regarding the virus and the process. After formulation, dosing should be done with the help of bulk antigen that helps to develop proposed vaccine candidate based on reference manual, and guidelines.

5. Cell culture procedure

CEF cell culture is widely used by researchers in the biopharmaceutical industry and veterinary vaccine production. The following steps should be maintaining chronologically to develop primary cell culture from the chicken embryo.

1. SPF eggs are incubated and collect the embryonated egg at 8–11 days (need base).

2. Cleaning the outside of egg by tincture of iodine, Cracking the egg into BSC.

3. Collect the embryo by forceps, place it into a sterile Petri dish, and chopping the body parts (embryo) with scissors.

4. Cut off head, wings, legs, remove the visceral parts and wash (PBS) rest of the body in Petri dish until clean the blood.

5. Chopping the clean body with scissors and gently pipette and aspirate by syringe.

6. Collect the suspension into a sterile falcon tube and perform mild centrifugation at 300 rpm for 5 min.

7. Add 0.25% trypsin EDTA to suspend the pellet by gentle pipetting (Recommendation: for 12 embryos add 10 ml trypsin and 10 ml PBS for suspension).

8. Centrifuge the suspension at 600 to 1000 rpm for 5 to 10 min (at 37°C).

9. Collect the supernatant into a new sterile falcon tube and filter it with double layer sterile gauze and collect it into a new tube and add 10 to 15 ml GM into it for 12 embryos. Generally, add GM two times (1:2) of collected cell/fluid.

10. Wash it by centrifugation at 10,000 rpm, 25°C for 10 min, and discard the supernatant. Then add GM to reconstitute and gently pipette. Repeat this step 2 times and collect the pellet (Suspected that 12 embryos produce 2 ml pellet).

11. Add 13 ml GM with the collected pellet (2 ml) and gently mix 15 ml cell suspension consider as a stock cell suspension.

12. Count the cell based on the stock with the help of a hemocytometer then split the incorporation of the cell with growth media that contain 10% FBS based on CCF measurement.

13. Observe the cell distribution within the CCF under an inverted microscope.

14. Place the CCF into a CO_2 incubator at 5% level, pH 7.2 to 7.4, 37°C for overnight. Then observe the cell morphology under a microscope.

6. Application of cell culture

Nowadays application of cell culture is exceptionally essential in life science and medical science. Cell culture technique is an excellent tool its applications are [1, 39, 40] as a). Production of pharmaceutical biochemicals b). Embryological study (CEF cell culture) c). Recombinant biomaterials (rDNA) and vaccine manufacturing, testing of the drug, drug sensitivity, and cytotoxicity of cell d). Production of human and animal vaccines (primary chicken fibroblast cells e). Manufacturing of immunotherapy f). Production of different enzymes, hormones (synthetic), immunobiological (like monoclonal antibodies, interleukins, lymphokines), and anticancer agents g). Cell culture is an excellent way to teach cell biology study h). Production of agricultural products like milk, (cultured) meat, fragrances i). In the microbiological study (Virus propagation, virology) j). Genetic engineers and biotechnologist are utilizing it within their field of research k). The aging, toxic compound study, cell morphology, cell physiology, and the study of mutagenesis cell culture have great impact.

7. Cell handling, storage and challenges

There are a lot of measures that should be taken to handle cells smoothly. Smooth and gentle handling of the cell gives better outcomes. An SOP must be defined clearly in the handling procedure of a cell. Clean area, BSL, pressure control of the cell culture room, personal hygiene, general laboratory management, gowning, skilled manpower can play a vital role in this regard. Generally, cell incubate at 36–37°C in a 5% CO_2 sometimes CO_2 level 4–10%, and time required based on cell types, research methodology. For the short preservation (few hours <2 hours) of the cell need normal freezing and for about day-long preservation required −20°C with growth media or cryoprotectant agent. Cryopreservation is required for the yearlong storage of cells. The temperature logbook is an essential document in the lab. Any fluctuation, problems in power supply, user entry of fridge/ cold room, and incubator are inevitably helpful to handle and maintain cells. Besides lock, access control or password system in laboratory, incubator, and fridge might be helpful in this regard. Resulting in easily identify the problems if any. Vigorous pipetting during cell harvesting, splitting cause cell damage sometimes causes cell death. A few toxic and harmful substances are eluted from the microfilters during sterilization. That's why practicing the sterile technique strictly and selecting cultural instruments carefully. It is recommended that washing all the instruments with the culture medium immediately before performing cell culture. Nevertheless, other types of contamination happened from plastic instruments or trace elements, even in water that affects the cells in culture [11]. Viruses, bacteria, mycoplasma, and endotoxins contamination may appear due to rough handling and haphazard performance. Strict environmental control is necessary to check cross-contamination in cell culture [41].

8. Quality control, GMP and cGMP

Quality control (QC) is known as maintaining the quality and authenticity of anything related to the research or laboratory work like cell, media, PBS, Polymerase chain reaction (PCR), or RT-PCR report. That's why the QC department, as well as personnel are important. Every life science researcher ought to know and aware of the current Good Manufacturing Practice (cGMP), QC process and system. QC, GMP, and cGMP have the same objectives and more or less same activities. In GMP and cGMP all are the followings clearly defined- Rules, regulations, and guidelines; good implementation; all aspects of the product's lifecycle and manufacturing process; product development and raw materials selection to the final production process; testing; storage, and shipment. The worlds policy maker organizations like FDA, Medicines & Healthcare Regulatory Agency (MHRA), WHO, European Union (EUDRALEX)-UK, approved all the guidelines of cGMP. Recently GMP is prescribed as "cGMP". The "c" stands for "current" as a reminder that all the techniques, systems, all the processes must be kept up-to-date to consent to the most recent regulations. [42]. cGMP guidelines ensure excellent quality at each step of work. It acts as a safeguard for the quality, purity, strength and identity of cells and related products. This guideline helps to construct a consolidated quality control and management system, helps to hold raw materials quality, helps to develop good SOPs, helps to a build a system of product quality investigation and deviation measurement, and take part in the building of reference laboratories for testing. The necessity of cGMPs are always keep in mind because it is very much important to establish QC. Numerous pharmaceutical companies are executing modern and comprehensive quality control systems and risk management system as a result they hold their quality above the average standards [43]. Quality control (QC) is employed in the GMP framework. Using a broad array of analytical techniques, the QA (Quality assurance) team will identify and quantify all the factors deemed critical for any specific product. Biological products require extensive analysis for characterization. There are some cations related to QC [44] such as-

- identity (confirms correct material);
- quantity (confirms dose parameter);

- purity (confirms material has correct purity);
- impurity (confirms product safety);
- potency (confirms activity);
- sterility including adventitious agents (confirms material is sterile).

9. Overall challenges: cell culture

Culturing techniques for CEF cells or BHK-21 or others is always challenging. The followings are the most important points by maintaining these points researchers or laboratory personnel easily can overcome any possible challenges. The points that need to pay attention such as a). Lab requirement is in the first, set up all the required materials, chemicals, instruments before the work or research b). Novice personnel should be avoided c). Skilled manpower and staff should be appointed for the work and as a trainer for new one d). Interpretation of work daily e). Establish a strong QC department f). Store all the true data and metadata because it is important to identify problems and also help to design a research g). Calibration of pH meter regular basis is mandatory and pH level follow up because cell health depends on it h). Need to take necessary precautions during culture media preparation i). Maintain and check the required CO_2 percentage based on cell j). Incubation time and temperature vary from cell to cell and it may be vary based on research work k). Transportation process, medium, time, precautions should be describing in transportation SOP l). Pay attention during thawing of the cell because it causes stress on the cell resulting cell might be death m). More careful and be cautious during preservation of cells, changing of media, passaging, trypsinization, harvesting of cells, mycoplasma contamination, and antibiotics treatment (for bacterial contamination) because these are the fundamental steps in cell culture technology. A non-technical or unskilled or novice personnel can easily ruin all the effective work at a time during the above points. That's why need to be more cautious about this n). Certificate of Analysis (COA) is essential for all the work like media preparation, pH level check, cell morphology, cell count, splitting, transportation, cell purchase or sell, and so on. It reduces the risk O). Based on international reference lab or manual SOP should be developed, SOP of how to prepare a good SOP and working guideline can minimize problems and it will take part to handle challenges. Current Good Manufacturing Practice (cGMP), Documentation of lab work, Training facility both for researchers and staff, Validation of work, calibration of instruments regularly can eradicate the problems and smoothly handle all the challenges.

10. Effectiveness of maintaining logbook, training and SOPs

Knowing the difference between a laboratory logbook and a laboratory notebook is very much effective to proceed with laboratory work. Logbook keeps details, usually in a tabulated format on handling and use of equipment, users, time of use, the purpose of use, and watched comments in case any. Other than, the research facility notebook or diary contains whole experimental details (for investigation and analysis), all the readings, results of any calculation, eventually all the supplementary data (graphs, spectra, or chromatograms) preserved in note book. Every research facility action related to the utilization of testing equipment, laboratory environmental records (temperature, humidity, atmospheric pressure, and exposure to light), weighing balance logs (analytical balance is the foremost used device in any research facility), material consumption, chemicals, details of suppliers and supplies are conserved as a proof record. Depending on the work nature vary the types of logbooks. Both notebooks and logs play a vital role in decision making or identify the problems or take part to fix the false result. During an audit, a logbook helps to convince the auditor because it provides evidence-based data. Periodically maintenance, operated by trained staff, calibration, and servicing of instruments increase the shelf life of the instrument, and these logs are stored as a proof document for safe work. Standard operating procedures (SOPs) based work has the same impact on research. On the other hand, training is significant for the advancement of the skills and knowledge of laboratory staff. In a training log the topic of training, date and time, name of trainer and trainees, effectiveness, and also the performance of individuals is reserved [34, 45]. Does every detail clarify in SOP like how to operate an instrument, test, or a machine? How will be eligible to perform? How to monitor (lab, work, test, etc.) and keep the record? How to calibrate and when? Who certifies and how? How to use chemicals? In a word SOP is fundamental. It is an ideal practice to have procedures for maintaining and controlling laboratory stocks which are clearly defined in specific SOP.

11. Conclusion

In recent years' importance and the application of cell culture cannot explain in a word. Cell culture technology is an extensively accepted technique in the field of molecular biology study and life science research. Due to high feasibility, cell culture practices highly demandable in biopharmaceutical works. This document summarizes the theoretical background, basic concepts regarding cell culture (animal cells, BHK-21, CEF cell culture). The above discussion gives an ideal concept to understand the whole process of cell culture technique, its applications and it helps to build constructive problem-solving confidence related to cell culture along with inspiration. Conclusively, it is said that all the steps regarding cell culture give a fundamental idea of cell culture.

Acknowledgements

Alhamdulillah. Author really thankful to his family members for *their* continuous support and encouragement.

Funding

Self-funded. This chapter did not receive any grant from funding agencies.

Conflict of interest

The author declares no conflict of interest.

Abbreviations

ATCC	American Type Culture Collection
BHK	Baby Hamster Kidney

BME	Basal Medium
BOD	Biological Oxygen Demand
BSC	Biosafety Cabinet
BSL	Biosafety Level
CCF	Cell Culture Flask
$CCID_{50}$	Cell Culture Infectious Dose 50
CDC	Center for Disease Control and Prevention
CEF	Chicken Embryo Fibroblast
cGMP	Current Good Manufacturing Practice
CHO	Chinese Hamster Ovary
COA	Certificate of Analysis
CPAs	Cryoprotective Agents
DAPI	4′,6-diamidino-2-phenylindole
DMEM	Dulbecco's Modified Eagle's Medium
DMSO	Dimethyl Sulphoxide
DT	Doubling Time (Population)
EC_{50}	Effective Concentration
EDTA	Ethylenediaminetetraacetic Acid
EID_{50}	Embryo Infective Dosage
ELD_{50}	Embryo Lethal Dosage
ELISA	Enzyme-Linked Immunosorbent Assay
EMEM	Eagle's Minimum Essential Medium
FBS	Fetal Bovine Serum
FDA	Food and Drug Administration
FFA	Focus forming Assay
FMD	Foot and Mouth Disease
GCCP	Good Cell Culture Practice
GM	Growth Media
GMP	Good Manufacturing Practice
HA	Hemagglutination Assay
hpi	hours post-inoculation
IC_{50}	Concentration of an Inhibitor
IMDM	Iscove's Modified Dulbecco's Medium
IPA	Isopropyl Alcohol
ISO	International Organization for Standardization
LD_{50}	Lethal Dose 50
MHRA	Medicines & Healthcare Regulatory Agency
MOI	Multiplicity of Infection
NIH	National Institute of Health, USA
OECD TG	Organization for Economic Co-Operation and Development Test Guide Line
PBS	Phosphate Buffer Saline
PCR	Polymerase Chain Reaction
PDL	Population Doubling Level
PFU	Plaque-Forming Unit
PPE	Personal Protective Equipment
QA	Quality Assurance
QC	Quality Control
R&D	Research and Development
rDNA	Recombinant Deoxyribonucleic Acid
RDS	Room Data Sheet
ROS	Reactive Oxygen Species
rpm	Rotation Per Minute

RT-PCR	Reverse Transcription Polymerase Chain Reaction
SOP	Standard Operating Procedures
SPF	Specific Pathogen Free
$TCID_{50}$	Tissue Culture Infectious Dose 50
TEM	Transmission Electron Microscopy
UPS	Uninterrupted Power Supply
WHO	World Health Organization

Author details

Md. Salauddin
Faculty of Veterinary, Animal and Biomedical Sciences, Department of Microbiology and Public Health, Khulna Agricultural University (KAU), Khulna, Bangladesh

*Address all correspondence to: salauddin.dvm@gmail.com; salauddin@kau.edu.bd

References

[1] Farzaneh M, Attari F, Mozdziak PE, Khoshnam SE. The evolution of chicken stem cell culture methods. British Poultry Science, 2017, 58 (6): 681-686. DOI:10.1080/00071668.2017.1365354

[2] Mandrioli M, Monti M, Tedeschi R. A practical guide to insect cell cultures: establishment and maintenance of primary cell cultures. HALTERES, 2015, 6: 132-141.

[3] Freshney RI. Culture of Animal Cells: A Manual of Basic Technique and Specialized Applications. 2010, Wiley, John & Sons, Inc, New Jersey.

[4] DGDA (Directorate General of Drug Administration). Guidelines on Animal Vaccine. 2018, Ministry of Health and Family Welfare Government of the People's Republic of Bangladesh.

[5] Kantardjieff A, Zhou W. Mammalian Cell Cultures for Biologics Manufacturing. In: Mammalian Cell Cultures for Biologics Manufacturing. 2014, Springer, Heidelberg, p. 1.

[6] Auniņš JG. Viral Vaccine Production in Cell Culture. In: Encyclopedia of Industrial Biotechnology: Bioprocess, Bioseparation, and Cell Technology. Wiley, New York 2010.

[7] Stoker M, Macpherson I. Syrian hamster fibroblast cell line Bhk21 and its derivatives. Nature, 1964, 203:1355-1357.

[8] Butler M. In: Animal Cell Culture and Techonology, Internl, T.J (ed.). Padstow, Cornwall, 2nd edition, UK, 2003, 10-27.

[9] Zhang, XZ, Chen Y, Huang HL, Xu DL, Ren CB, Liu BT, Su S, Tang ZX. Pseudorabies virus induces viability changes and oxidative stress in swine testis cell-line. Pakistan veterinary journal, 2013, 33: 438-441.

[10] Griffiths JB. Mammalian cell culture reactors, scale-up. In: Flickinger MC (ed) Encyclopedia of industrial biotechnology: bioprocess, bioseparation, and cell technology. Wiley, New York, 2010, 1-13.

[11] Yao T, Asayama Y. Animal-cell Culture media: History, characteristics, and current issues. Reprod Medical Biology. 2017; 16: 99-117. https://doi.org/10.1002/rmb2.12024

[12] Brunner D, Frank J, Appl H, Schoffl H, Pfaller W, Gstraunthaler G. Serum-free cell culture: the serum-free media interactive online database. ALTEX, 2010, 27: 53-62.

[13] Bhatia S, Naved T, Sardana S. Introduction to Pharmaceutical Biotechnology, Volume 3 Animal tissue culture and biopharmaceuticals. Chapter 1. Introduction to animal tissue culture science. IOP Publishing Ltd, 2019. doi:10.1088/2053-2563/aafac0ch1

[14] Verma A, Verma M, Singh A. Animal tissue culture principles and applications. Animal biotechnology: tools and techniques. Chapter 14. Elsevier Inc.2020, 269-293.https://doi.org/10.1016/B978-0-12-811710-1.00012-4

[15] Nema R, Khare S. An animal cell culture: Advance technology for modern research. Advances in Bioscience and Biotechnology. 2012; 3: 219-226.

[16] Pohjanpelto P. Putrescine transport is greatly increased in human fibroblasts initiated to proliferate. The Journal of cell biology, 1976; 68: 512-520.

[17] Ibtisham F, Zhao Y, Nawab A, Liguang H, Wu J, Xiao M, Zhao Z, An L. The Effect of High Temperature on Viability, Proliferation, Apoptosis and Anti-oxidant Status of Chicken Embryonic Fibroblast Cells. Brazilian

Journal of Poultry Science, 2018, 20 (3): 463-470.

[18] Dubey AK, Lavanya L, Sadananda D, Gouthami K, Elfansu K, Singh A. Inferences of Carbon Dioxide in Present-Day Cell Culture Systems: An Unacknowledged Problem and Perspectives. Austin Therapeutics, 2021, 6(1): 1033.

[19] Maria TL, Eric KJ, Ehrhardt RO. Cryopreservation and Thawing of Mammalian Cells. In: eLS. John Wiley & Sons, Ltd: Chichester, 2014. DOI: 10.1002/9780470015902.a0002561.pub2

[20] Choi J, Bischof JC. Cooling rate dependent biophysical and viability response shift with attachment state in human dermal fibroblast cells. Cryobiology, 2011, 63: 285-291.

[21] Hunt CJ, Armitage SE, Pegg DE. Cryopreservation of umbilical cord blood: 2. Tolerance of CD34+ cells to multimolar dimethyl sulphoxide and the effect of cooling rate on the recovery after freezing and thawing. Cryobiology, 2003, 46: 76-87.

[22] Naaldijk Y, Staude M, Fedorova V, Stolzing A. Effect of different freezing rates during cryopreservation of rat mesenchymal stem cells using combinations of hydroxyethyl starch and dimethylsulfoxide. BMC Biotechnology, 2012, 12: 49.

[23] Ragai RM, Robin DH. Human Cell Culture Protocols, Methods in Molecular Biology, vol. 806, DOI 10.1007/978-1-61779-367-7_1

[24] ATCC. ANIMAL CELL Animal Cell Culture Guide. www.atcc.org

[25] Wayne M, Maria YL. Rolf TO. Ehrhardt. Cryopreservation and Thawing of Cells. Current Protocols in Immunology. Wiley Online Library. John Wiley & Sons, Inc.2012. DOI: 10.1002/0471142735.ima03gs99

[26] Mather JP, Roberts PE. Introduction to Cell and Tissue Culture: Theory and Technique. 1998.

[27] Bhatia S, Naved T, Sardana S. Introduction to Pharmaceutical Biotechnology, Volume 3 Animal tissue culture and biopharmaceuticals. Chapter 1. Introduction to animal tissue culture science. IOP Publishing Ltd, 2019. doi:10.1088/2053-2563/aafac0ch1

[28] Science Gateway. Cell Biology Protocols. https://www.sciencegateway.org/protocols/cellbio/cell/moipfu.htm

[29] Kärber G. Beitrag zur kollektiven Behandlung pharmakologischer Reihenversuche. Archiv f. Experiment. Pathol. U. Pharmakol. Springer-Verlag, 1931, 162 (4): 480-483. doi:10.1007/BF01863914

[30] Kaufmann SH, Kabelitz D. Methods in Microbiology: Immunology of Infection. Academic Press. 2002, vol-32. ISBN 0-12-521532-0.

[31] Liang T, Liu X, Cui S, Qu S, Wang D, Liu N, Wang F, Ning K, Zhang B, Zhang D. Generation of a reliable full-length cDNA of infectious Tembusu virus using a PCR-based protocol. Virus Res. 2016, 213, 255-259.

[32] Shahla Shahsavandi, Mohammad Majid Ebrahimi, Ashraf Mohammadi, Nima Zarrin Lebas. Impact of chicken-origin cells on adaptation of a low pathogenic influenza virus. Cytotechnology, 2013, 65:419-424. DOI 10.1007/s10616-012-9495-5

[33] Coecke S. Guidance on good cell culture practice: A Report of the Second ECVAM Task Force on good cell culture practice. ATLA Alternatives to Laboratory Animals, 2005,33 (3): 261-287.

[34] OECD. Test No. 432: In Vitro 3T3 NRU Phototoxicity Test, OECD Guidelines for the Testing of Chemicals,

Section 4, OECD Publishing, Paris, 2004, http://dx.doi.org/10.1787/9789264071162-en

[35] Maneesha S, Inamdar, Lyn Healy, Sinha A, Stacey G. Global Solutions to the Challenges of Setting up and Managing a Stem Cell Laboratory. Stem Cell Rev and Rep, 2012, 8:830-843. DOI 10.1007/s12015-011-9326-7

[36] Coecke S, Balls M, Bowe G, Davis J, Gstraunthaler G, Hartung T. Guidance on good cell culture practice (GCCP). A report of the second ECVAM task force on good cell culture practice. ATLA, 2005, 33: 1-27.

[37] Drexler HG, Upho CC. Mycoplasma Contamination of Cell Cultures: Incidence, Sources, Effects, Detection, Elimination, Prevention. Cytotechnology, 2002, 39: 75.

[38] RegMedNet. Biological Industries. 2017. https://www.regmednet.com/10-tips-for-preventing-mycoplasma-contamination/

[39] Maguire G. Therapeutics from Adult Stem Cells and the Hype Curve. ACS Medicinal Chemistry Letters, 2016, 7 (5): 441 3. doi:10.1021/acsmedchemlett.6b00125.

[40] Prieto D, Aparicio G, Sotelo-Silveira JR. Cell migration analysis: A low-costlaboratory experiment for cell and developmental biology courses using keratocytes from fish scales. Biochemistry and Molecular Biology Education, 2017, 45 (6):475-482. doi: 10.1002/bmb.21071

[41] Ragai R, Mitry, Robin D, Hughes. Human Cell Culture Protocols, Methods in Molecular Biology, vol. 806, DOI 10.1007/978-1-61779-367-7_1

[42] BI (Biological Industries). cGMP: GeneralCell Culture, Stem Cell Research. An Interview with Ravid Grimberg. 2019. https://blog.bioind.com/cgmp-what-every-life-science-researcher-should-know-about-current-good-manufacturing-practice/

[43] FDA. Facts About the Current Good Manufacturing Practices (CGMPs) https://www.fda.gov/drugs/pharmaceutical-quality-resources/facts-about-current-good-manufacturing-practices-cgmps

[44] Green A, Sharpe G. Good Manufacturing Practice for Cell Culture Processing. Medicines from Animal Cell Culture Edited by G. Stacey and J. Davis. John Wiley & Sons, Ltd, 2007. ISBN: 978-0-470-85094-7

[45] OECD. OECD Principles on Good Laboratory Practice, OECD Series on Principles of Good Laboratory Practice and Compliance Monitoring, No. 1, OECD Publishing, Paris 1998. http://dx.doi.org/10.1787/9789264078536-en.

Chapter 2

Overview of Primary Cell Culture Models in Preclinical Research of Prostate and Bladder Cancer

Kalyani Killekar, Sridevi I. Puranik, Aimen Akbar A., Shridhar C. Ghagane, Rajendra B. Nerli and Murigendra B. Hiremath

Abstract

The number of patients diagnosed with prostate and bladder cancer is increasing worldwide and one of the most important challenges remains the development of effective, safe and economically viable antitumor drugs. Clinical approval for drugs tested in preclinical studies enabling them to enter phase I clinical trials is essential. Cell lines are *in vitro* model systems that are widely used in different fields of medical research, especially basic cancer research and drug discovery. Their usefulness is primarily linked to their ability to provide an indefinite source of biological material for experimental purposes. Under the right conditions and with appropriate controls, authenticated cancer cell lines retain most of the genetic properties of the cancer of origin. Studies conducted during the initial development of drugs such as toxicity, corrosion and drug activity were carried out on animals; however, in the past two decades, alternatives have been sought due to the fact that animals do not effectively model to human *in vivo* conditions and unexpected responses are observed in the studies. Also, more than 100 million animals were used and billion dollars were spent for animal toxicity experiments. Cell culture studies made positive contributions to the initial development of drugs and is highly desirable, as it provides systems for ready, direct access and evaluation of tissues. Contrary to animal studies, less cost and the need for low drug and a short response time are the characteristics for *in vitro* cell culture methods. *In vitro* tumor models are a necessary tool, in not only the search for new substances showing antitumor activity but additionally for assessing their effectiveness. This chapter reviews the main features of primary cancer cell cultures, provides an overview of the different methods for their selection and management, and summarizes the wide range of studies that can be performed with them to improve the understanding of prostate and bladder cancer preclinical treatment processes.

Keywords: Primary cell culture, preclinical studies, prostate and bladder cancer, *in vitro* model

1. Introduction

Cancer is the one of the major death cause worldwide and accounted nearly 10 million deaths in 2020 [1]. The rate of incidence in prostate cancer and bladder cancer are increasing worldwide too. According to GLOBOCON 2018, prostate cancer is the second most frequent cancer and in men, it is fifth leading cause of death. Bladder cancer is also common in men ranking on sixth position and ninth leading cause of cancer death [2]. There are so many treatments available like radiotherapy, chemotherapy, hormonal therapy but these treatments are associated with adverse side effects and poor quality of post treatment life. Hence there is need in development of effective, safe and economically viable antitumor drugs.

Prostate cancer and bladder cancer are heterogeneous diseases where many molecular, environmental and genetic factors are involved in its progression and understanding the mechanism of this progression is difficult [3]. In recent years the cancer research has made significant progress, but many challenges remain as it is [4]. Currently, only 7% of potential anticancer drugs are gaining approval which is much lower than drugs for other diseases [5]. Hence, to improve this percentage, it is essential to clinically approve drugs which are tested in preclinical studies and enabling them to enter phase I clinical trials [6].

Experimental models are important tools in the cancer research. The model should be reproducible, able to successfully reflect disease stage that is being studied and mimic the disease; how it behaves in humans [4]. Cell lines are *in vitro* model systems, a necessary tool, in not only the search for new substances showing antitumor activity but additionally for assessing their effectiveness. They are widely used in different fields of medical research and pharmaceutical companies. Presently pharmaceutical industries mostly rely on *in vitro* models like two dimensional (2D), three dimensional (3D), boyden's chamber (to study chemotaxis and assessment of cell motility) [7], micro fluidic systems (It is small devices that can provide a specific fluid flow, constant temperature, fresh medium, flow pressure and chemical gradients which is same as *in vivo* systems to study migration and invasion [8], 3D bioprinting (mimics the processes that occurs in tumor micro environment) [9, 10]. Main reason for accepting *in vitro* model is it's physiological relevance, it helps in improving the understanding of prognosis and treatment, it provides accuracy and it is also a low cost screening tool for researchers [11]. The usefulness of *in vitro* models is primarily linked to their ability to provide an indefinite source of biological material for experimental purposes. The *in vivo* model involves animals which provide valuable information to understand many aspects in development of disease and initial development of drugs such as toxicity, corrosion and drug activity [12]. But from past two decades, alternatives have been sought due to the fact that animals do not effectively model humans in *in vivo* conditions, as it shows unexpected responses like anatomical variation and also difficulty in extracting quantitative mechanistic data in the studies. Mathematical models are also used in the cancer research to analyze tumor growth and progression, and helps in predicting the effects of some therapies [13]. Different clinical setting, cancer resistance and switching to another treatment, existence of unknown biological details these issues can affect the mathematical models [14–16]. Computer simulation is another model in the cancer research, helps to test complex multi scale cancer progresses, it also accounts for drug pharmacokinetics and pharmacodynamics, but has drawback in less common cancers because of less data, therefore it lacks perspective validation and accuracy [17].

All models involved in the cancer research have pros and cons hence the cost duration, experimental design and data analysis in developing the anticancer drug should be considered for the selection of the model. It is necessary to choose more effective preclinical platforms to screen the antitumor compounds [18]. Practically *in-vitro* models of tumors will not only give primary screening of potential anti-tumor drugs but it also prevents drugs with insufficient antitumor activity from entering into preclinical animal testing [19]. This chapter reviews the main features of primary cancer cell cultures, provides an overview of the different methods for their selection and management, and summarizes the wide range of studies that can be performed with them to improve the understanding of prostate and bladder cancer preclinical treatment processes.

2. Primary cell culture

Primary cell culture is a gold standard testing platform for *in-vitro* research in oncology as they reflect the tumor state more accurately compare to most commonly employed cell lines [20]. It is a powerful tool commonly used by scientists to study cellular properties and mechanisms of isolated cells in a controlled environment [21].

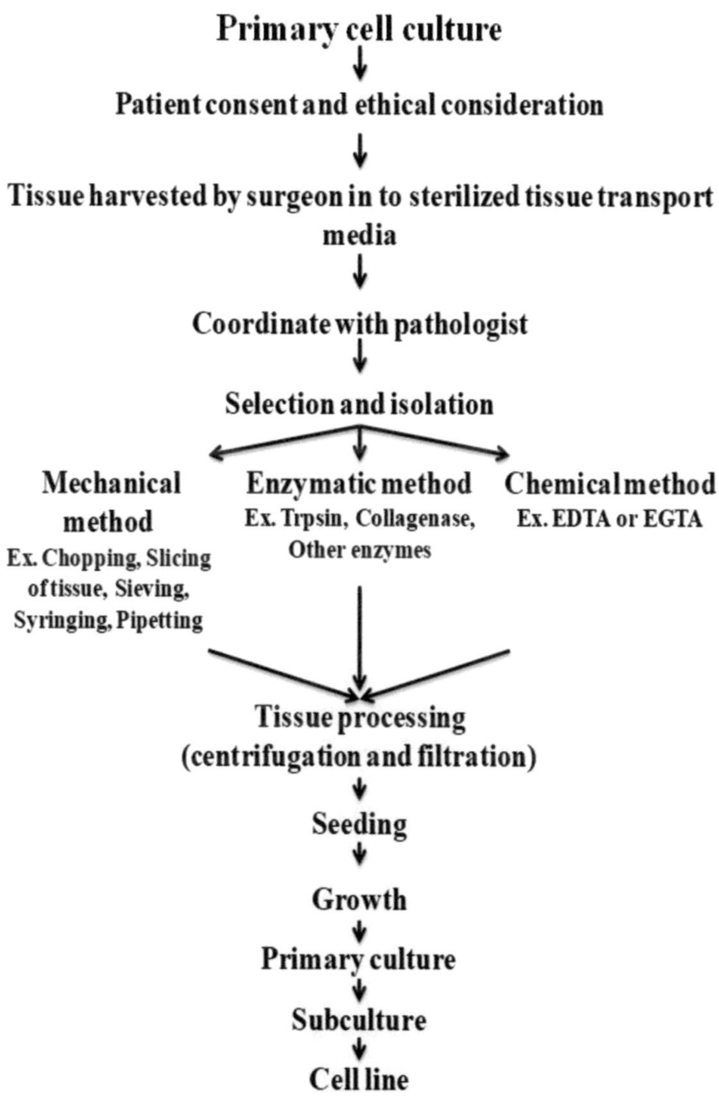

Figure 1.
Flowchart showing the process of primary cell culture technique.

Cell culture studies have made positive contributions to the initialdevelopment of drugs. Contrary to animal studies *in vitro* method requires low drug dose and short response time, which is characteristic feature of in vitro cell culture methods [22]. Primary cell culture is also called as *ex-vivo*. Because primary cells are directly taken from tissue origin and cultured under favorable conditions hence it is more similar to the *in-vivo* state and exhibit normal physiology. It maintains the cross talk between malignant and healthy components [23]. This is the main reason why primary cell culture is called as excellent model system to carry studies in metabolic, aging, signaling studies and effects of drugs, toxic compounds in the cell.

Primary cells are non-immortalized and non-transformable as they imitate the appearance of living model and hence these can help to model 3D tissues. In this culture, cells will grow in 3D aggregates and presents interesting application [24]. It helps for detection, isolation, growth and developmental stages of viruses and helps to analyze the mode of infection. Drug candidate and its toxicity screening rely on results of early-stage *in vitro* cell based assays. Particularly in pharmaceutical industry primary cell culture is used to synthesize verity of biomolecules in high scale, various research project on cell-based product are developed. It is alternative for animal model to test effect of drugs and cosmetics [25]. There are few technical hurdles associate with primary cell culture. For instance, culturing might be difficult if the quality of the surgical material is poor. Also, due to early onset of cell senescence, difficulty arises to maintain sufficient number of cell passages but researchers have made many attempts to resolve this problem [26, 27].

Primary cell culture has been subdivided in to adherent cell culture and suspension cell culture. In the adherent cell culture, cells are arranged in monolayer and attach to the surface of the culture flask. Adherent cells are usually derived from tissues of organs. Growth is limited to surface area and it needs to be detached from the surface before cells get sub-cultured. Viral vaccine, gene therapy and cell therapies depend only on adherent cell culture. Suspension cells are derived from the peripheral blood and do not require any attachment for growth. They do not get attached the surface area of the cultural flask. The cells are free floating and growth is limited to the concentration of cells. The steps involved in the primary cell culture are represented in (**Figure 1**).

3. Isolation of cells

Before going to any further tissue processes, it is important to keep in mind that all tissue processing has to be carried out in a biosafety cabinet and all the sterilization protocols has to be maintained properly [28]. Now moving towards cell isolation, it is a process where one or more specific cells are isolated from heterogenous cell mixture. Isolation of primary cells from cancer cells is an important phenomenon of cell culture biology as they are more reliable sources to understand the human cell. There are many standard protocols available for culturing the normal andneoplastic cells [28]. Human prostate and bladder are composed of many cell types which can be isolated and cultured. Hierarchy of the epithelium has been reviewed most [29, 30]. There are 3 main epithelial lineages namely neuroendocrine, basal and luminal. Prostatic homeostasis is mainly depends upon the epithelial cells and stromal cells; stromal cells guides to the epithelium cell for their dedifferentiation, proliferation and also progression of carcinogenesis [31].

Now, how to understand which cell is cancerous and which cell is non- cancerous because cell does not contain tags on it. There are specific cell markers (Antigen) which will identify the difference between cancerous and non-cancerous for ex. ARA70 (Androgen receptor-associated protein 70) is a cell marker which was noted

Cell marker	Cell type
Cytokeratin 5 [33]	Basal cell
CD59 [34]	Basal cell
c-Met [35]	Basal cell
CD95 [34]	Basal cell
Cytokeratin 8 [36]	luminal cell
CD9 [37]	luminal cell
15-LOX-2 [36]	luminal cell
CD24 [34]	luminal cell

Table 1.
Biomarkers for Prostate cell culture.

Cell marker	Cell type
CD9 [38]	Urothelial
CD104+ [38]	Basal
CD13+ [38]	Stromal cells of the lamina propria
CK5 [37]	Basal
p63 [37]	Basal
CK8 [37]	Luminal

Table 2.
Biomarkers for bladder cell culture.

Antigen	Antibodies	Cell type
Cytokeratin 8/18 (Prostate)	Mouse IgG1	Luminal cells and intermediate cells
High molecular weight cytokeratin (Prostate)	Mouse IgG1	Intermediate cells
Trop2-APC (Prostate)	Mouse IgG2a	Epithelial cells
CD49f-PE(Prostate)	Rat IgG2a	Epithelial cells
CD + 9 (Bladder)	antibody clone M-L13	Urothelial
CD104 (Bladder)	clone 439-9B	Basal

Table 3.
Primary antibodies for prostate and bladder epithelial cultures [38, 40].

to be expressed at high levels in normal primary cultures compared with prostate cancer cell lines [32]. Many cell markers are available depending upon cell type which is listed in **Table 1**. Also in bladder cancer depending upon cell type there are different CD (Cluster of Differentiation) cell markers which are depicted in **Table 2** [38]. Identification of stem cell marker has uncovered a cellular hierarchy of epithelium during development and in response to injury [39]. Cell markers (Antigens) have specific antibodies and these have to be evaluated histochemically. These reactions are evaluated by specific kits which are available in market (**Table 3**). For isolation, first tissues will be collected from prostate cancer patients and bladder

Method	Mechanism	Advantages	Disadvantages
Chemical Method	By using EDTA or EGTA it binds with cations and disrupt the intracellular bonds [42]	Easy and cost saving	It does not adequately dissociate all types of tissue
Mechanical Method	Cutting, scratching, the tissues in to small pieces in order to separate the cells and wash it with gentle agitation [42]	It is a rapid technique works best for loosely associated tissue. Correct temperature should be maintained for enzyme.	Decreases in the surviving capacity of the cell, incision of scissor, scalpel for cutting, scratching can damage the cell.
Enzymatic Method	Enzymes to cut or digest tissue pieces in free cells. Combination of enzyme also can be used Ex.- Collagenase, Trypsin, Hyluronidase [43].	It has great specificity with specific enzymes	Enzyme dissociation can modify proteins on cell surface

Table 4.
Mechanisms for isolation of cells.

Sl. No.	Components	Function
1	Buffering systems 1. Natural buffer system 2. Chemical buffers system	• CO_2 balances the pH,5 to 10% Co_2 incubation, non-toxic and cost effective [44] • HEPES, buffering capacity7.2–7.4, in high concentration it is toxic, costly [45, 46]
2	Inorganic salt	Maintains osmotic balance and membrane potential by providing sodium, potassium, and calcium ions [47]
3	Amino acids	Require for proliferation of cells and provides nitrogen Ex. L-glutamine provides NAD,NADPH and serve as secondary source of energy in metabolism [48].
4	Carbohydrates	Carbohydrates are major source of energy, most of the media use glucose and Galactose.
5	Proteins and peptides	Major proteins and peptides in media are albumin, fibronectin and tranferrin. Albumin helps to remove toxic substances from cell culture media
6	Fatty acids and lipids	Particularly added when serum free media is used [44].
7	Vitamins	Important for cell growth and proliferation as cells cannot produce sufficient amount, need to provide through culture media [44].
8	Trace elements	Copper, Zink, Selenium required in trace amount for proper growth and many biological process
9	Antibiotics	Controls growth of bacteria and fungi ex.Penicillin, Streptomycin, fluconazole [49].
10	Serum in media	Serum is a complex mixture provides all above elements. Ex. Fetal and Calf bovine serum [44].

Table 5.
Basic components of media and their functions.

cancer patients who are undergoing biopsy. This collection of tissues needs to be well coordinated between urology, pathology and the investigator and it has to proceed for primary culture within 2 hours after collection of tissues [33–37, 39–41].

After collection, tissue should be placed in sterile container which has HBSS (Hanks' balanced salt solution) with HEPES (Hydroxyethyl piperazineethanesulfonic acid) and store at 4°C for 2 hours to increase cell viability. To asses tumor cells in the dissected material Hematoxylin and Eosin (H&E) stain is used in the histopathology lab. To get a single cell suspension from tissue dissociation obtained after sugary, there are three mechanisms available for isolation: Chemical, Mechanical and Enzymatic method (**Table 4**).

Although this is first step in primary culture, there is still no standardized protocol for this. There is a variety of options available. Tissue has to be mechanically minced from autoclaved scalpel or scissor; if tissue is measuring from 1 to 20 grams semi-automated dissociator can be used. Manual method has to be done in ice cold PBS (Phosphate-buffered saline). Commercially available formulation showed 10% increased viability compare to collagenase I, II, IV. (**Table 5**) [33]. In another study mechanical and enzymatic method has been used. In mechanical method, they used lacerate and scalpels and in enzymatic method collagenase type I and hyaluronidase type I enzymes with medium agitation at 37°C for 18 hours was used [34]. EDTA (Ethylenediaminetetraacetic acid)/Trypsin mixture used with 5 minutes of incubation in 37°C degree for prostate tissue [35]. Both the mechanisms, mechanical disaggregation with disposable disaggregator and enzymatic by collagenase and trypsin used for prostate tissue [36, 37]. In some cases, trypsin/EDTA 1:5 solution and incubation for 15 minutes for bladder tissue was used. In some studies for dissociation of bladder tissue 1:1 collagenase II and dispase enzymes are used at 37°C for 12 hours [40–43]. Also there is a need to monitor tissue digestion process for every 2 hours by gently shaking the digestion mixture by checking the viability of cells under the microscope [50–53].

4. Tissue processing

After digestion, cells are strained by strainer to separate the debris from it. Then, cells micro clumps are washed with PBS or HBSS twice or tricefollowed by centrifugation [54]. Cell pellet collected from centrifugation is suspended in 2 ml of culture media. Count the viable cell by hemocytometer or by tryptophan dye exclusion method [55]. Cell viability also can be measured by the intracellular adenosine triphosphate levels which are commercially available kit [56]. Immunohistochemistry and immunofluorescence techniques used to localize, identify and quantitate the cells based on cell surface marker [57].

5. Culture media

Cell pellets collected from centrifugation has to be placed in micro well plate or flask that contains culture media. It provides artificial environment for cell to grow. Basic requirement of culture media are controlled temperature, substrate to attach cell, growth medium and incubator to maintain pH [45]. Main step in culture is to choose culture media. It generally composed of amino acid, vitamins, inorganic salts, glucose, hormones, growth factor, and attachment factor which provides energy and helps to complete the cell cycle. Commercially available cell media for primary epithelialcancer cells are less effective compare to tissue specific primary cell media prepared in lab (**Table 6**) [46–49].

Choice of culture media is very important to get significant result in experiment. Selection of media completely depends upon type of cell, purpose and resource [62]. As primary culture provides valuable research data, preparation of quality

SL No.	Cell type	Components of media
1.	Fibroblast cell culture [28]	DMEM media with 7.2 pH + Fetal calf serum(FCS) + 100 U/ml penicillin,100 µg/ml streptomycin +1% amphotericin B added in culture media.culture plate incubated at 37°C a humidified chamber of 95% air and 5% CO_2
2.	Prostate cancer (Bone metastatic variant) [29]	DMEM Glutamax +4.5 g/L D-Glucose with pyruvate +10%FCII +1% penicillin–streptomycin+37°C with 5% CO_2.
3.	Bladder cancer (Epithelial cells) [58]	EMEM (ATCC) +10% FCS +1% penicillin + streptomycin + humidified incubator at 37°C with 5% CO_2.
4.	Prostate cancer (Epithelial cells) [59]	KSFM medium+ + 25 mM HEPES +1% penicillin + streptomycin +0.5 mg/mL fungizone +100 mg/mL gentamicin +37°C, 5% CO 2 humidified incubation.
5.	Epithelial cells [40]	serum-free RPMI 1640 without phenol red+ penicillin 100 IU/mL+ streptomycin 100 µg/mL+ metronidazole 1 µg/mL+ amphotericin B 2.5 µg/mL + gentamicin 20 µg/mL + 37°C and 5% CO_2 for 6 days.
6.	Bladder cancer (Urothelial cells) [50]	Glutamine + insulin + Phosphoethanolamine+ ethanolamine + hydrocortisone+ transferrin + EGF + BPHE+5% FBS +5% CO_2 at 37°C
7.	Prostate cancer (Epithelial cells) [60]	DMEM + Glucose +100 U/ml penicillin +100 mg/ml streptomycin sulfate+0.29 mg/ml glutamine + Euroclone
8.	Prostate cancer (Epithelial cells) [61]	DEME /Ham's (1:1) + BSA (0.01%) + FAS (2%) + Epidermal growth factor (10 ng/mL) + (insulin-transferring-selenium-1%) + hydrocortison (0.5 µg/mL) + Tryiodotyronone(1 nM) + phosphpethenolamine(0.1 mM) + choleratoxin (50 ng/mL) + fibronectin(100 ng/mL) + futine(20 µg/mL) + penicillin/streptomycin (100 U/mL,100 µg/mL) + R1881(0.1 nM).

Table 6.
Components of media from different studies.

Sl.No.	Name of media	Supplier
1.	Human Endothelial-SFM	Life Technologies
2.	Endothelial Basal Media	Sigma Aldrich
3.	EndoGRO-LS Complete Media Kit	MilliporeSigma
4.	HUVEC Basal Medium CB HUVEC	AllCells
5.	Endothelial Cell Medium	ScienCell
6.	Epithelial cell medium	ScienCell
7.	EpiGRO primary epithelial cells	MilliporeSigma
8.	RPMI 1640	Sigma Aldrich
9.	αMEM	ThermoFischer
10.	k- SFM	ThermoFischer

Table 7.
Commercially available media for epithelial cells [44, 63].

culture media is required, or to avoid limitation (cell number) of primary cell culture, commercially produced medias are available (**Table 7**) [58–61].

It is very important to maintain cell viability after isolation process which is totally depends on skillful handling and culture conditions. The culture condition will differ depending on the cell type. Cell growth has to be observed till 11 or 12 days. Additional extra media, Fetal Bovine Serum (FBS) and antibiotics need to be provided to avoid contamination. Culture media has to be changed between 2 and 3 days [60]. Initially apoptosis is 5% from 0 to 1 day but as days will pass apoptosis rate will increases from 7 to 14 days. But functional validity of benign and prostate cancer cells was 5 days after confirming it with histochemically, biochemical and by immunohistochemical assay [63]. Use of serum free culture media with low calcium condition increases the longevity of the cell. Cryopreservation (Preservation of structurally intact cells) can be achieved by adding 10% FBS (Fetal bovine serum) and 10% DMSO (Dimethyl sulfoxide) in 80% confluence primary cell culture [64].

Each day morphological changes have to benoted. Normal cells get counted every day and cancer cells get counted every 2 days [65]. Cell viability is determined by trypsin blue dye, equal volume of PBS and trypsin blue dye allowed to sit on cells for few minutes then to count the cells samples are loaded on hemocytometer, cells scored as leaving or dead based on uptake of tryptophan blue dye [66]. Once confluence reaches to 80 to 90% it has to get counted by Neubauer camera at 1:2 dilution with tryptophan dye exclusion, MTT assay (3-(4,5-dimethylthiazol-2-yl)-2,5-diphenyl tetrazolium bromide) can be used to determine cell viability [67]. Cell growth curve can be plotted from the graph to check the time when the cell viability increases in the culture. Once cell get cultured properly depending upon need of investigator, cells can be passaged and characterization of the cells can be done.

6. Cancer cell lines

Most established prostate cancer cell lines namely PC-3, DU145 (Duke University 145) and LNCaP (Lymph Node Carcinoma of the Prostate) developed in the 1970s and 1980s are in the majority of the published studies [68–70]. T24 cells (cell line from transitional cell carcinoma of the bladder) are exemplified for bladder cancer research [71]. These immortalized cancer cell lines are not always predictive of the real cancer behavior for the preclinical studies as these cells are adapted to 2D monolayer culture conditions [72, 73].

7. Cell culture models

Traditionally animal models were commonly employed for carrying out study of different types of cancer for the past three decades [74]. These animal studies have many drawbacks including lack of high-throughput drug screening, longer time consumption to conduct tests and ethical controversies concerning animal testing. Cell culture is the most widely used alternative to animal studies and cell culture techniques can broadly be classified into 2D and 3D methods [75]. The potentialities of primary cancer cell models' cultures in preclinical studies for cancer research and drug discovery has amplified over the past few years. Primary cell cultures provide a good model system to understand normal and malignant biological activities. Carcinogenesis-related behavior such as apoptosis, proliferation, adhesion, differentiation, migration, senescence, invasion, angiogenesis, and other metabolic pathways have been studied in recent years. One of the major advantages is that the heterogeneity of cell populations composing a primary culture mimics the tumor microenvironment, crosstalk, and interactions between malignant and healthy cells, neither of which is possible with cell lines [76].

Most studies have shown that the cellular responses to drug treatments in 3D cell culture are significant and more similar to that of *in vivo* architecture when compared to 2Dcell culture. One of the most improved successful assays using 3D culture for cell-based screening in the early phase of drug discovery is cancer cell viability assessment. This assay is particularly useful to test the cytotoxic effects of compounds that may lead to cell death. It plays an essential role in checking how many cells are viable at the end of each experiment. Cell viability assay is closely followed by cell proliferation, cell migration and then cell signaling assays [77–79].

Currently a number of anticancer drugs belonging to different classes chemically are available. To be used as a potential anticancer agent, the testing compounds need to inhibit the growth and proliferation of cancer cell lines. This will further inhibit the signaling pathways by knocking in or knocking out a candidate gene thereby stopping the progression of tumor to fatal stages. For instance, antiproliferative investigations were performed on prostate cancer cell line DU-145 *in vitro* and *in vivo* using salvia miltiorrhizabunge [80]. Another example to justify this concept will be a study performed on PC-3 cell line. Generally, cancer cells express higher amount of Transferrin Receptors (TfR) for an increased uptake of iron in relative to normal cells. This higher amount of intracellular free iron is required for the growth and proliferation of cancer cells. Anhydrodihydroartemisinin (ADHA) was used to inhibit PC-3 cell lines through caspase-dependent pathway [81].

8. Applications: cell culture in drug discovery and screening

8.1 Drug candidate identification

Often the rate-limiting step in preclinical drug discovery is the target identification and validation step. 3D cell cultures have the potential to discover the molecular perturbations governing carcinogenesis and to accelerate target identification and validation, given that the gene expression patterns found in 3D models are relative to in vivo, when compared to 2D monolayer models [81]. For instance, a study reported that CXCR7 (C-X-C chemokine receptor type 7) and CXCR4 (C-X-C chemokine receptor type 4) were co-expressed in LNCaP, DU145 and PC3 cell lines in 2D culture. A marked up-regulation of both receptors was observed in PC3 cells when cultured in 3D using Matrigel suggesting that inhibition of CXCR7/CXCR4 may assist in controlling prostate tumor growth and subsequent progression [82].

8.2 Toxicity profiling

Cultured cancer cells are powerful in assessing drug-induced toxicity and to determine suitable drugs and methods for selectively destroying different types of cancer. It is useful to investigate effects of drug responses on metabolic signaling pathways or candidate genes conceding drug screening practices with impressive progress in the last decade. A study investigated features such as vascularization

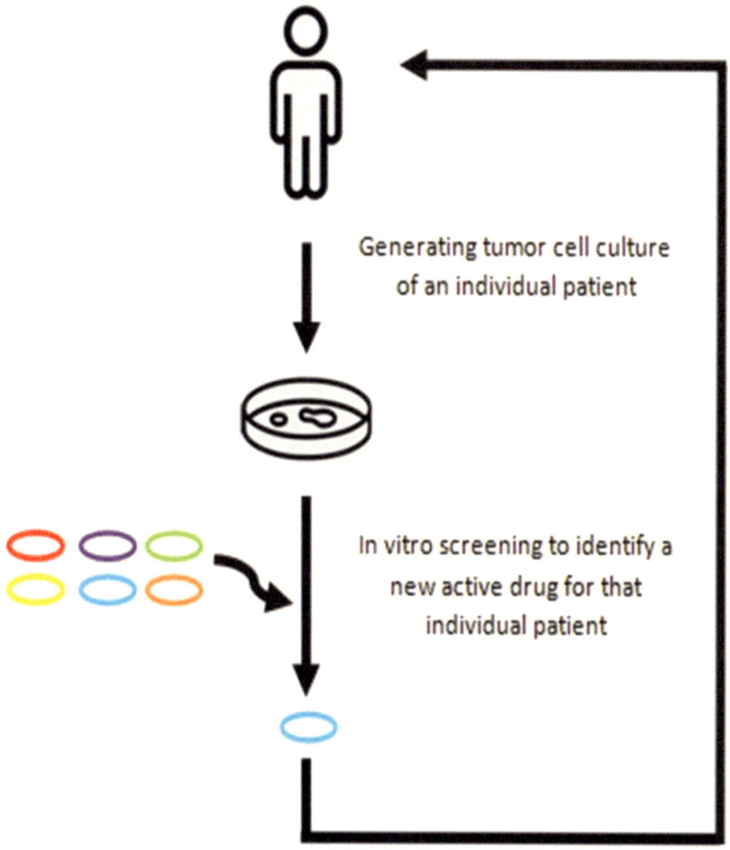

Figure 2.
Assay-guided treatment choosing the best active drug for an individual patient.

and perfusion of antineoplastic drugs on human T24 bladder cancer [83]. It allowed in the understanding of basic paracrine signaling mechanisms that regulates tissue homeostasis, development of new methods for urinary bladder reconstruction and tissue engineering, and generation of models of malignant and benign diseases. This study suggested that the use of 3D urinary bladder cultures could be a possible approach in clinical practice to select for the best antineoplastic drug for each patient and to investigate the effect of drug combinations or new antineoplastic drugs [84, 85]. The below (**Figure 2**) suggests how assay-guided treatment can be useful in choosing the best active drug for an individual patient.

8.3 Testing anticancer activities

By far the most useful *in vitro* model which is used to analyze the anticancer activity is Cell culture. Treatments including radiotherapy, chemotherapy, hormone therapy, novel and experimental therapies can be evaluated. Extracts of plants can also be utilized to check for anti-cancer behaviors such anti-inflammatory, destabilized membranes through which invasion and migration can occur. For instance, leaf extracts of *Leea indica* were used to study *in– vitro* antioxidant and anticancer activity on DU-145 and PC-3 human prostate cancer cell lines [86]. An example of drug combination is the synergistic effect of cisplatin and sunitnib malate – based chemotherapy on T24, 5637, and HT1376 human urinary bladder cancer cell lines [87].

9. Conclusion

Primary cell cultures have its application in various fields like toxicology, virology, drug screening, genetic engineering, gene therapy, genetic counseling, cancer research but main important application is model system. It provides best model system for studying basic cell biology and biochemistry, effects of disease-causing agent and cell, effect of drugs on cell, process which triggers aging and apoptosis. Primary cell culture represents excellent model for transitional preclinical experiments to understand cancer in *in-vitro* system. Primary cell culture acts as gold standard for cell line experiments because it provides broader spectrum of cell types from greater number of patients to be studied without any induction of artificial genetic mutation and it also maintains same phenotype throughout the culture. It involves both, clinician and researcher in the culture, it helps in understanding the drawback of treatments and lack in laboratory methodology and hence it is possible to overcome from it.

There are many models in cancer research; each model has different potentialities and inadequacy. In primary cell culture, complexity arises due to poor tissue quality, collection and inappropriate culturing may decrease the cell viability. Hence management of primary cell culture is difficult. But to overcome from these difficulties proper collection with the help of pathologist and selection of proper isolation method and culture media based on tissue type can help to increase the cell viability. Considering the current clinical system towards precision medicine, patient derived cancer models are powerful epitome in cancer research. Nowadays 3D model system is emerging system. Primary cell culture can help to model 3D culture, in future technological perspectives like 3D culture can replace the *in vivo* model system. In conclusion, this chapter reviews several aspects of primary cell culture, provides overview on selection of tissues, different methods of isolation, culturing media and management of cells after culture. It summarizes the wide range of studies to improve the understanding of prostate and bladder cancer preclinical treatment processes.

Author's contributions

KK & SIP: Worked on collection of data and drafted the chapter, AAA: Worked on the draft of cell lines images and edited, SCG: Developed the study designedand edited the chapter, RBN & MBH: Collected the literature and guided throughout the study. All the authors reviewed and approved the chapter.

Funding

Nil.

Conflict of interest

The authors declare conflict of interest as none.

Abbreviations

ADHA	Anhydrodihydroartemisinin
BPHE	Brazed Plate Heat Exchangers

BSA	Bovine Serum Albumin
CD cell markers	Cluster of Differentiation cell markers
DMEM	Dulbecco's Modified Eagle Medium
DMSO	Dimethyl sulfoxide
DU145	Duke University 145
EDTA	Ethylenediaminetetraacetic acid
EGTA	Ethylene glycol tetraacetic acid
EMEM	Eagle's Minimum Essential Medium
FBS	Fetal Bovine Serum
FCS	Fetal Calf Serum
FGF	Fibroblast Growth Factors
HBBS	Hanks' balanced salt solution
HEPES	Hydroxyethylpiperazineethanesulfonic acid
IgMI	Imunoglobuline M Imunostain
KSFM	Keratinocyte Serum Free Medium
LNCaP	Lymph Node Carcinoma of the Prostate
PBS	Phosphate buffer saline
RPMI 1640	Roswell Park Memorial Institute (*RPMI*) *media*
TfR	Transferrin Receptors

Author details

Kalyani Killekar[1], Sridevi I. Puranik[2], Aimen Akbar A.[3], Shridhar C. Ghagane[4,5]*, Rajendra B. Nerli[5] and Murigendra B. Hiremath[6]

1 Department of Biochemistry, JN Medical College, KLE Academy of Higher Education and Research (Deemed-to-be-University), Belagavi, Karnataka, India

2 Department of Zoology, KLES B. K. Arts, Science and Commerce College, Chikodi, Karnataka, India

3 Department of Parasitology, McGill University, Montreal, Quebec, Canada

4 Division of Urologic-Oncology, Department of Urology, KLES Dr. Prabhakar Kore Hospital and Medical Research Centre, Belagavi, Karnataka, India

5 Department of Urology, JN Medical College, KLE Academy of Higher Education and Research (Deemed-to-be-University), Belagavi, Karnataka, India

6 Department of Biotechnology and Microbiology, Karnatak University, Dharwad, Karnataka, India

*Address all correspondence to: shridhar.kleskf@gmail.com

References

[1] Cancer [Internet]. https://www.who.int/news-room/fact-sheets/detail/cancer

[2] Ferlay J, Colombet M, Soerjomataram I, Mathers C, Parkin DM, Piñeros M, Znaor A, Bray F. Estimating the global cancer incidence and mortality in 2018: GLOBOCAN sources and methods. International journal of cancer. 2019; 144(8):1941-1953.

[3] Rhim JS. In vitro human cell culture models for the study of prostate cancer. Prostate Cancer and Prostatic Diseases. 2000; 3(4):229-235.

[4] Nascimento-Gonçalves E, Ferreira R, Oliveira PA, Colaço BJ. An overview of current alternative models for use in the context of prostate cancer research. Alternatives to Laboratory Animals. 2020; 48(2):58-69.

[5] Solovyeva VV, Kitaeva KV, Rutland CS, Rizvanov AA. Cell Culture Based in vitro Test Systems for Anticancer Drug Screening. Frontiers in Bioengineering and Biotechnology.;8.

[6] Loucera C, Esteban-Medina M, Rian K, Falco MM, Dopazo J, Peña-Chilet M. Drug repurposing for COVID-19 using machine learning and mechanistic models of signal transduction circuits related to SARS-CoV-2 infection. Signal transduction and targeted therapy. 2020; 5(1):1-3.

[7] Falasca M, Raimondi C, Maffucci T. Boyden chamber. InCell Migration 2011 (pp. 87-95). Humana Press.

[8] Ruzycka M, Cimpan MR, Rios-Mondragon I, Grudzinski IP. Microfluidics for studying metastatic patterns of lung cancer. Journal of nanobiotechnology. 2019; 17(1):1-30.

[9] Kang HW, Lee SJ, Ko IK, Kengla C, Yoo JJ, Atala A. A 3D bioprinting system to produce human-scale tissue constructs with structural integrity. Nature biotechnology. 2016; 34(3):312-319.

[10] Truong D, Fiorelli R, Barrientos ES, Melendez EL, Sanai N, Mehta S, Nikkhah M. A three-dimensional (3D) organotypic microfluidic model for glioma stem cells–Vascular interactions. Biomaterials. 2019; 198:63-77.

[11] Katt ME, Placone AL, Wong AD, Xu ZS, Searson PC. In vitro tumor models: advantages, disadvantages, variables, and selecting the right platform. Frontiers in bioengineering and biotechnology. 2016; 4:12.

[12] C Gonçalves V, JLL Pinheiro D, de la Rosa T, G de Almeida AC, A Scorza F, A Scorza C. Propolis as a potential disease-modifying strategy in Parkinson's disease: cardioprotective and neuroprotective effects in the 6-OHDA rat model. Nutrients. 2020; 12(6):1551.

[13] Zazoua A, Wang W. Analysis of mathematical model of prostate cancer with androgen deprivation therapy. Communications in Nonlinear Science and Numerical Simulation. 2019; 66:41-60.

[14] Blackard CE, Byar DP, Jordan WP. Vacurg (1973) Orchiectomy for advanced prostatic carcinoma. Urology.;1(6):553.

[15] Hirata Y, Bruchovsky N, Aihara K. Development of a mathematical model that predicts the outcome of hormone therapy for prostate cancer. Journal of theoretical biology. 2010; 264(2): 517-527.

[16] Rich JN. Cancer stem cells in radiation resistance. Cancer research. 2007; 67(19):8980-8984.

[17] Pai RK, Van Booven DJ, Parmar M, Lokeshwar SD, Shah K, Ramasamy R,

Arora H. A review of current advancements and limitations of artificial intelligence in genitourinary cancers. American Journal of Clinical and Experimental Urology. 2020;8(5):152.

[18] Imamura Y, Mukohara T, Shimono Y, Funakoshi Y, Chayahara N, Toyoda M, Kiyota N, Takao S, Kono S, Nakatsura T, Minami H. Comparison of 2D-and 3D-culture models as drug-testing platforms in breast cancer. Oncology reports. 2015; 33(4):1837-1843.

[19] Stevens JL, Baker TK. The future of drug safety testing: expanding the view and narrowing the focus. Drug discovery today. 2009; 14(3-4):162-167.

[20] Janik K, Popeda M, Peciak J, Rosiak K, Smolarz M, Treda C, Rieske P, Stoczynska-Fidelus E, Ksiazkiewicz M. Efficient and simple approach to in vitro culture of primary epithelial cancer cells. Bioscience reports. 2016;36(6): e00423.

[21] Ager-Wick E, Hodne K, Fontaine R, von Krogh K, Haug TM, Weltzien FA. Preparation of a high-quality primary cell culture from fish pituitaries. JoVE (Journal of Visualized Experiments). 2018 (138): e58159.

[22] ŞAHİN ŞH, MESUT B, ÖZSOY Y. Applications of Cell Culture Studies in Pharmaceutical Technology. ACTA Pharmaceutica Sciencia.;55(3).

[23] Yin L, Qin G, Qian HZ, Zhu Y, Hu W, Zhang L, Chen K, Wang Y, Liu S, Zhou F, Xing H. Continued spread of HIV among injecting drug users in southern Sichuan Province, China. Harm reduction journal. 2007; 4(1):1-7.

[24] Joseph JS, Malindisa ST, Ntwasa M. Two-dimensional (2D) and three-dimensional (3D) cell culturing in drug discovery. Cell Culture. 2018; 2:1-22.

[25] https://bioscience.lonza.com/lonza_bs/IN/en/what-are-primary-cells-and-how-do-they-differ-from-cell-lines [Internet] 2021

[26] Stoczynska-Fidelus E, Piaskowski S, Bienkowski M, Banaszczyk M, Hulas-Bigoszewska K, Winiecka-Klimek M, Radomiak-Zaluska A, Och W, Borowiec M, Zieba J, Treda C. The failure in the stabilization of glioblastoma-derived cell lines: spontaneous in vitro senescence as the main culprit. PloS one. 2014; 9(1): e87136.

[27] Zieba J, Ksiazkiewcz M, Janik K, Banaszczyk M, Peciak J, Piaskowski S, Lipinski M, Olczak M, Stoczynska-Fidelus E, Rieske P. Sensitivity of neoplastic cells to senescence unveiled under standard cell culture conditions. Anticancer research. 2015; 35(5):2759-2768.

[28] Ranganathan K, Kavitha L. Oral epithelial dysplasia: Classifications and clinical relevance in risk assessment of oral potentially malignant disorders. Journal of oral and maxillofacial pathology: JOMFP. 2019; 23(1):19.

[29] Shen MM, Abate-Shen C. Molecular genetics of prostate cancer: new prospects for old challenges. Genes & development. 2010; 24(18):1967-2000.

[30] Wang Y, Hayward SW, Cao M, Thayer KA, Cunha GR. Cell differentiation lineage in the prostate. Differentiation. 2001 Oct 1;68(4-5):270-279.

[31] Taylor RA, Risbridger GP. Prostatic tumor stroma: a key player in cancer progression. Current cancer drug targets. 2008; 8(6):490-497.

[32] Tekur S, Lau KM, Long J, Burnstein K, Ho SM. Expression of RFG/ELE1α/ARA70 in normal and malignant prostatic epithelial cell

cultures and lines: regulation by methylation and sex steroids. Molecular Carcinogenesis: Published in cooperation with the University of Texas MD Anderson Cancer Center. 2001; 30(1):1-3.

[33] Brawer MK, Peehl DM, Stamey TA, Bostwick DG. Keratin immune reactivity in the benign and neoplastic human prostate. Cancer research. 1985; 45(8):3663-3667.

[34] Liu AY, Peehl DM. Characterization of cultured human prostatic epithelial cells by cluster designation antigen expression. Cell and tissue research. 2001 Sep;305(3):389-397.

[35] You X, Yu HM, Cohen-Gould L, Cao B, Symons M, Woude GF, Knudsen BS. Regulation of migration of primary prostate epithelial cells by secreted factors from prostate stromal cells. Experimental cell research. 2003; 288(2):246-256.

[36] Van Leenders G, Dijkman H, Hulsbergen-van de Kaa C, Ruiter D, Schalken J. Demonstration of intermediate cells during human prostate epithelial differentiation in situ and in vitro using triple-staining confocal scanning microscopy. Laboratory investigation. 2000; 80(8):1251-1258.

[37] Sfakianos JP, Daza J, Hu Y, Anastos H, Bryant G, Bareja R, Badani KK, Galsky MD, Elemento O, Faltas BM, Mulholland DJ. Epithelial plasticity can generate multi-lineage phenotypes in human and murine bladder cancers. Nature communications. 2020;11(1):1-6.

[38] Liu AY, Vêncio RZ, Page LS, Ho ME, Loprieno MA, True LD. Bladder expression of CD cell surface antigens and cell-type-specific transcriptomes. Cell and tissue research. 2012; 348(3):589-600.

[39] Strand DW, Aaron L, Henry G, Franco OE, Simon W. populations. 2017;91(224):139-51. DOI10.1016/j. diff.2015.10.013.Isolation

[40] Failli A, Consolini R, Legitimo A, Spisni R, Castagna M, Romanini A, Crimaldi G, Miccoli P. The challenge of culturing human colorectal tumor cells: establishment of a cell culture model by the comparison of different methodological approaches. Tumori Journal. 2009; 95(3):343-347.

[41] Caspar A, Mostertz J, Leymann M, Ziegler P, Evert K, Evert M, Zimmermann U, Brandenburg LO, Burchardt M, Stope MB. In vitro cultivation of primary prostate cancer cells alters the molecular biomarker pattern. in vivo. 2016; 30(5):573-579.

[42] Castell JV, Gómez-Lechón MJ. Liver cell culture techniques. Hepatocyte transplantation. 2009:35-46.

[43] Chen S, Principessa L, Isaacs JT. Human prostate cancer initiating cells isolated directly from localized cancer do not form prostaspheres in primary culture. The Prostate. 2012; 72(13): 1478-1489.

[44] Meenakshi A. Cell culture media: a review. Mater Methods. 2013; 3:175.

[45] Shipman Jr C. Evaluation of 4-(2-hydroxyethyl)-1-piperazineëthanesulfonic acid (HEPES) as a tissue culture buffer. Proceedings of the Society for experimental biology and medicine. 1969; 130(1):305-310.

[46] Zigler JS, Lepe-Zuniga JL, Vistica B, Gery I. Analysis of the cytotoxic effects of light-exposed HEPES-containing culture medium. In Vitro Cellular & Developmental Biology. 1985; 21(5):282-287.

[47] Činátl J. Inorganic-organic multimolecular complexes of salt solutions, culture media and biological fluids and their possible significance for

the origin of life. Journal of theoretical biology. 1969; 23(1):1-0.

[48] Pasieka AE, Morgan JF. Glutamine metabolism of normal and malignant cells cultivated in synthetic media. Nature. 1959; 183(4669):1201-1202.

[49] Perlman D. Use of antibiotics in cell culture media. InMethods in enzymology 1979 Jan 1 (Vol. 58, pp. 110-116). Academic Press.

[50] Roberson KM, Yancey DR, Padilla-Nash H, Edwards DW, Nash W, Jacobs S, Padilla GM, Larchian WA, Robertson CN. Isolation and characterization of a novel human bladder cancer cell line: BK10. In Vitro Cellular & Developmental Biology-Animal. 1998; 34(7):537-544.

[51] Goldstein AS, Drake JM, Burnes DL, Finley DS, Zhang H, Reiter RE, Huang J, Witte ON. Purification and direct transformation of epithelial progenitor cells from primary human prostate. Nature protocols. 2011; 6(5):656-667.

[52] Grainger A. Difficulties in tracking the long-term global trend in tropical forest area. Proceedings of the National Academy of Sciences. 2008; 105(2):818-823.

[53] Navone NM, Olive M, Troncoso P. Isolation and culture of prostate cancer cell lines. InCancer Cell Culture 2004 (pp. 121-132). Humana Press.

[54] Kim FJ, Campagna A, Khandrika L, Koul S, Byun SS, Bokhoven AV, Moore EE, Koul H. Individualized medicine for renal cell carcinoma: establishment of primary cell line culture from surgical specimens. Journal of endourology. 2008; 22(10):2361-2366.

[55] Ali MY, Anand SV, Tangella K, Ramkumar D, Saif TA. Isolation of primary human colon tumor cells from surgical tissues and culturing them directly on soft elastic substrates for traction cytometry. Journal of visualized experiments: JoVE. 2015(100).

[56] Sharma S, Neale MH, Di Nicolantonio F, Knight LA, Whitehouse PA, Mercer SJ, Higgins BR, Lamont A, Osborne R, Hindley AC, Kurbacher CM. Outcome of ATP-based tumor chemosensitivity assay directed chemotherapy in heavily pre-treated recurrent ovarian carcinoma. BMC cancer. 2003; 3(1):1-0.

[57] Maity B, Sheff D, Fisher RA. Immunostaining: detection of signaling protein location in tissues, cells and subcellular compartments. Methods in cell biology. 2013; 113:81-105.

[58] van de Merbel AF, van der Horst G, van der Mark MH, van Uhm JI, van Gennep EJ, Kloen P, Beimers L, Pelger R, van der Pluijm G. An ex vivo tissue culture model for the assessment of individualized drug responses in prostate and bladder cancer. Frontiers in oncology. 2018; 8:400.

[59] Sato T, Clevers H. Primary mouse small intestinal epithelial cell cultures. InEpithelial cell culture protocols 2012 (pp. 319-328). Humana Press, Totowa, NJ.

[60] Papini S, Rosellini A, De Matteis A, Campani D, Selli C, Caporali A, Bettuzzi S, Revoltella RP. Establishment of an organotypic in vitro culture system and its relevance to the characterization of human prostate epithelial cancer cells and their stromal interactions. Pathology-Research and Practice. 2007; 203(4):209-216.

[61] Zhang W, van Weerden WM, de Ridder CM, Erkens-Schulze S, Schönfeld E, Meijer TG, Kanaar R, van Gent DC, Nonnekens J. Ex vivo treatment of prostate tumor tissue recapitulates in vivo therapy response. The Prostate. 2019; 79(4):390-402.

[62] Clifford WJ, Anellis A, Ross Jr EW. Evaluation of media, time and

temperature of incubation, and method of enumeration of several strains of Clostridium perfringens spores. Applied microbiology. 1974; 27(4):784-792.

[63] Wang R, Chu GC, Wang X, Wu JB, Hu P, Multani AS, Pathak S, Zhau HE, Chung LW. Establishment and characterization of a prostate cancer cell line from a prostatectomy specimen for the study of cellular interaction. International journal of cancer. 2019;145(8):2249-2259.

[64] Maund SL, Nolley R, Peehl DM. Optimization and comprehensive characterization of a faithful tissue culture model of the benign and malignant human prostate. Laboratory Investigation. 2014; 94(2):208-221.

[65] Speirs V, Green AR, Walton DS, Kerin MJ, Fox JN, Carleton PJ, Desai SB, Atkin SL. Short-term primary culture of epithelial cells derived from human breast tumours. British journal of cancer. 1998; 78(11):1421-1429.

[66] Samsel L, Zaidel G, Drumgoole HM, Jelovac D, Drachenberg C, Rhee JG, Brodie AM, Bielawska A, Smyth MJ. The ceramide analog, B13, induces apoptosis in prostate cancer cell lines and inhibits tumor growth in prostate cancer xenografts. The Prostate. 2004; 58(4):382-393.

[67] Raxworthy MJ. Animal cell culture: A practical approach: Edited by RI Freshney. pp 248. IRL Press, Oxford. 1986.£ 22 or£ 14 (pbk) ISBN 0-947946-62-4 or 0-947946-33-0 (pbk).

[68] Horoszewicz JS. The LNCaP cell line-a new model for studies on human prostatic carcinoma. Prog Clin Biol Res. 1980; 37:115-132.

[69] Mickey DD, Stone KR, Wunderli H, Mickey GH, Vollmer RT, Paulson DF. Heterotransplantation of a human prostatic adenocarcinoma cell line in nude mice. Cancer research. 1977; 37(11):4049-4058.

[70] Kaighn ME, Narayan KS, Ohnuki Y, Lechner JF, Jones LW. Establishment and characterization of a human prostatic carcinoma cell line (PC-3). Investigative urology. 1979;17(1):16-23.

[71] Bubenik J, Barešová M, Viklický V, Jakoubkova J, Sainerova H, Donner J. Established cell line of urinary bladder carcinoma (T24) containing tumour-specific antigen. International journal of cancer. 1973; 11(3):765-773.

[72] Hamacher R, Bauer S. Preclinical models for translational sarcoma research. Current opinion in oncology. 2017; 29(4):275-285.

[73] Gillet JP, Varma S, Gottesman MM. The clinical relevance of cancer cell lines. Journal of the National Cancer Institute. 2013; 105(7):452-458.

[74] Ravi M, Paramesh V, Kaviya SR, Anuradha E, Solomon FP. 3D cell culture systems: advantages and applications. Journal of cellular physiology. 2015; 230(1):16-26.

[75] Joseph JS, Malindisa ST, Ntwasa M. Two-dimensional (2D) and three-dimensional (3D) cell culturing in drug discovery. Cell Culture. 2018; 2:1-22.

[76] Miserocchi G, Mercatali L, Liverani C, De Vita A, Spadazzi C, Pieri F, Bongiovanni A, Recine F, Amadori D, Ibrahim T. Management and potentialities of primary cancer cultures in preclinical and translational studies. Journal of translational medicine. 2017;15(1):1-6.

[77] Comley J. 3D cell culture, easier said than done. Drug Discov World. 2010:25-41.

[78] Riss TL, Moravec RA, Niles AL, Duellman S, Benink HA, Worzella TJ, Minor L. Cell viability assays. Assay Guidance Manual [Internet]. Eli Lilly & Company and the National Center for Advancing Translational Sciences. 2016.

[79] Bae WJ, Choi JB, Kim KS, Ha US, Hong SH, Lee JY, Hwang TK, Hwang SY, Wang ZP, Kim SW. Inhibition of Proliferation of Prostate Cancer Cell Line DU-145 in vitro and in vivo Using Salvia miltiorrhiza Bunge. Chinese journal of integrative medicine. 2020; 26(7):533-538.

[80] Ahmad F, Sarder A, Gour R, Karna SK, Arora P, Kartha KR, Pokharel YR. Inhibition of prostate cancer cell line (PC-3) by anhydrodihydroartemisinin (ADHA) through caspase-dependent pathway. EXCLI journal. 2020; 19:613.

[81] Ghosh S, Spagnoli GC, Martin I, Ploegert S, Demougin P, Heberer M, Reschner A. Three-dimensional culture of melanoma cells profoundly affects gene expression profile: A high density oligonucleotide array study. Journal of cellular physiology. 2005; 204(2):522-531.

[82] Lovitt CJ, Shelper TB, Avery VM. Advanced cell culture techniques for cancer drug discovery. Biology. 2014; 3(2):345-367.

[83] Zips D, Thames HD, Baumann M. New anticancer agents: in vitro and in vivo evaluation. in vivo. 2005; 19(1):1-7.

[84] Varley CL, Southgate J. Organotypic and 3D reconstructed cultures of the human bladder and urinary tract. In3D Cell Culture 2011 (pp. 197-211). Humana Press.

[85] Johannsen M, Gneveckow U, Thiesen B, Taymoorian K, Cho CH, Waldöfner N, Scholz R, Jordan A, Loening SA, Wust P. Thermotherapy of prostate cancer using magnetic nanoparticles: feasibility, imaging, and three-dimensional temperature distribution. European urology. 2007; 52(6):1653-1662.

[86] Ghagane SC, Puranik SI, Kumbar VM, Nerli RB, Jalalpure SS, Hiremath MB, Neelagund S, Aladakatti R. In vitro antioxidant and anticancer activity of Leea indica leaf extracts on human prostate cancer cell lines. Integrative medicine research. 2017; 6(1):79-87.

[87] Arantes-Rodrigues R, Pinto-Leite R, Fidalgo-Gonçalves L, Palmeira C, Santos L, Colaço A, Oliveira P. Synergistic effect between cisplatin and sunitinib malate on human urinary bladder-cancer cell lines. BioMed research international. 2013; 2013.

Chapter 3

Two-Dimensional and Three-Dimensional Cell Culture and Their Applications

Sangeeta Ballav, Ankita Jaywant Deshmukh, Shafina Siddiqui, Jyotirmoi Aich and Soumya Basu

Abstract

Cell culture is one of the most important and commonly used *in vitro* tools to comprehend various aspects of cells or tissues of a living body such as cell biology, tissue morphology, mechanism of diseases, cell signaling, drug action, cancer research and also finds its great importance in preclinical trials of various drugs. There are two major types of cell cultures that are most commonly used- two-dimensional (2D) and three-dimensional culture (3D). The former has been used since the 1900s, owing to its simplicity and low-cost maintenance as it forms a monolayer, while the latter being the advanced version and currently most worked upon. This chapter intends to provide the true meaning and significance to both cultures. It starts by making a clear distinction between the two and proceeds further to discuss their different applications *in vitro*. The significance of 2D culture is projected through different assays and therapeutic treatment to understand cell motility and treatment of diseases, whereas 3D culture includes different models and spheroid structures consisting of multiple layers of cells, and puts a light on its use in drug discovery and development. The chapter is concluded with a detailed account of the production of therapeutic proteins by the use of cells.

Keywords: Cell culture, 2D culture, 3D culture, drug action, therapeutics

1. Introduction

The growth of cells in a controlled artificial environment isolated from their natural habitat is referred to as cell culture [1]. It is a significant tool used widely to study cell and molecular biology, screening drugs and toxicity analysis, the role of a particular gene in a disease, and cancer research. Due to their unique properties, they also have been tuned for screening and developing biopharmaceutical compounds such as vaccines and recombinant proteins. One of the major advantages of using cell culture is the homogenous and reproducible data generated [2].

Drug discovery is a lengthy and time-consuming process that undergoes several stages of testing and optimization. This encompasses identification of the target, lead discovery, pre-clinical validation, and clinical trials [3]. Therefore, it is very pertinent to obtain information about the biological activity, biochemical mechanisms, toxicity, and off-target interactions of drug molecules leading to the early stages of drug discovery. *In vitro*, cell-based assays prove futile to understand the effects of drugs on the cells at an early stage of drug discovery which attributes an increased chance of development of drugs with good efficacy and safety [4].

Two-dimensional (2D) cell culture was introduced many decades ago that has been the major type of cell culture technique in numerous fields. This traditional approach has been extensively used for drug screening due to its relatively inexpensive feature and convenience to use. However, the issue of mimicking the *in vivo* environment restricts its use [5]. The 2D cell cultures grow as a monolayer in controlled flat environments, such as a glass or polystyrene flask that comprises live proliferating cells because of the detachment of dead cells from the surface (**Figure 1**). As a result, this leads to uniformity in nutrients and growth factors present in the medium to which the cells get access and proliferate at a faster rate than they would *in vivo* [6]. Thus, the morphology of the cells is completely changed as they appear flattered and stretched as compared to the *in vivo* environment. Besides this, the cell–cell interactions and cell-extracellular interactions become different in comparison to the tumor [7].

Recently there has been an upsurge of interest towards three-dimensional (3D) cell culture in biomedical research and drug development processes due to its high-throughput accuracy and refined *in vitro* models [8]. They have been broadly used in understanding the cell shape, cell–cell interaction, and the cellular environment that efficiently mimics the *in vivo* environment. 3D cultures grow as clusters or aggregates called spheroids either with a matrix or without a matrix [9] (**Figure 1**). There is a gradient of nutrients across these spheroids due to which the cells at the surface of spheroids proliferate more as compared to the cells that are present in the interior [10]. As a result of the difference in the proliferation rate, the cells in the spheroids are usually in different stages of the cell cycle such as proliferating, quiescent, hypoxic, and necrotic cells. In the 3D culture, the cells have uniform access to nutrients as in the case of a tumor. Also, the shape of the cells, cell–cell interactions, and cell-environment interactions are well defined in 3D culture [7]. It has also been observed that 2D cell cultures are more sensitive to drugs as compared to 3D cells [11]. This chapter intends to provide the true meaning and significance of both cultures. It starts by making a clear distinction between the two and proceeds further to discuss their

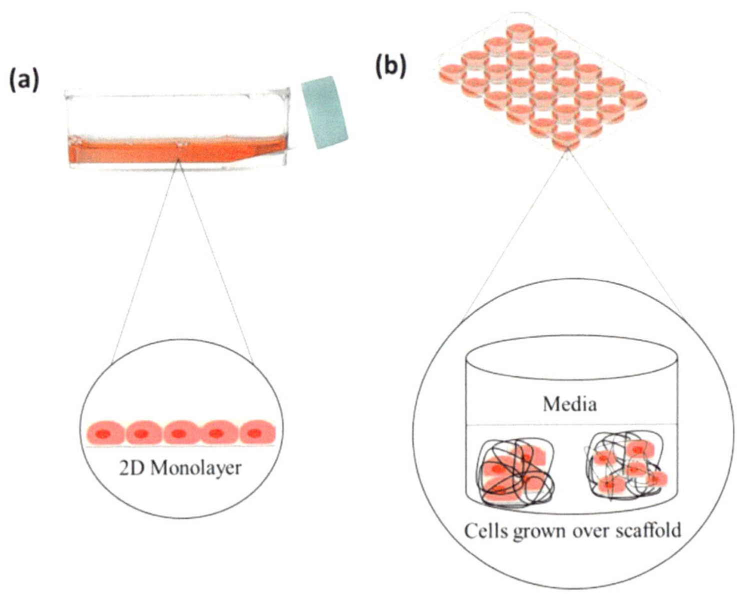

Figure 1.
Simplified sketch of 2D cell culture (a) and 3D cell culture (b).

different applications *in vitro*. The significance of 2D culture is projected through different assays and therapeutic treatment to understand cell motility and treatment of diseases, while 3D culture includes different models and spheroid structures consisting of multiple layers of cells, and puts a light on its use in drug discovery and development. The chapter is concluded with a detailed account of the production of therapeutic proteins by the use of cells.

2. *In vitro* applications of 2D culture

2.1 *In vitro* cytotoxicity assays and tissue-engineered tissue models

Cytotoxicity assays are commonly used for *in vitro* toxicology and pharmacology studies for the screening the effect of chemicals and drugs on the cultured cells. There are different assays available for measuring cytotoxicity namely- the colony-forming assay and dye inclusion or exclusion such as neutral red and trypan blue assay is the most significantly used. Cytotoxicity assays can be broadly divided into the following categories– (i) Assays based on metabolism, (ii) Adenosine triphosphate (ATP) Bioluminescence Assay and (iii) Assays based on the release of enzymes.

Assays based on metabolism generally include the 3-(4,5-dimethythiazol2-yl)-2,5-diphenyl tetrazolium bromide (MTT) assay and its alternatives such as 2,3-Bis-(2-Methoxy-4-Nitro-5-Sulfophenyl)-2H-Tetrazolium-5-Carboxanilide (XTT),3-(4,5-dimethylthiazol-2-yl)-5-(3-carboxymethoxyphenyl)-2-(4-sulfophenyl)-2H-tetrazolium(MTS) and sulforhodamine B (SRB) assay. Due to rapid, quantitative, versatile, and highly reproducibility of MTT, it is widely used in large-scale, anti-tumor drug-screening program. MTT is a quantitative colorimetric assay that quantifies the reduction of yellow tetrazolium dye by mitochondrial succinate dehydrogenase to purple insoluble formazan crystals by the NADPH dependent cellular oxidoreductase enzymes [12]. The crystals are dissolved in an appropriate solvent. The absorbance is then recorded using a spectrophotometer to analyze the cell viability wherein the crystals get accumulated in the viable cells due to their impermeability to the cell membrane.

ATP Bioluminescence Assay is used to measure the ATP level that is well regulated in the metabolically active live eukaryotic cells as compared to the dead cells wherein the ATP level falls due to the activity of ATPases. This assay includes a luciferase enzyme that utilizes energy from ATP that converts luciferin into oxyluciferin and thus produces luminescence. Therefore, luminescence could be used to measure the ATP level. Assays based on the release of enzymes are more significant as they measure the products released by the dead cells [13].

Assays based on the release of enzymes include Lactate dehydrogenase(LDH) leakage assay involving the formation of pyruvate from lactate in the presence of LDH with simultaneous reduction of NAD to NADH that alters the absorbance at 340 nm [14]. Research in cancer and cell biology is greatly dependent on *in vitro* assays and models. This help in understanding the various responses of the cultured cells when exposed to different conditions. Tissue-engineered *in vitro* tissue models serve as an alternative to *in vivo* animal studies to study the physiology of various diseases. Example of tissue-engineered *in vitro* models includes Skeletal Muscle Models [15], blood–brain barrier model [16], aneurysm models [17] and the Pre-vascularized Human Vaginal Mucosa model [18].

2.2 Cell migration assay

Cell migration is well known for its significant role in embryonic morphogenesis, cancer invasion and metastasis, immune responses, tissue formation, and angiogenesis [19]. Mainly, cell migration is of two types; single-cell migration and collective cell migration. Single-cell migration is regulated by cytoskeletal activity without cell-to-cell interactions with neighboring cells. This type of migration is important for embryonic development, immune response, and in the early stages of metastasis. On the other hand in the collective cell migration, the group of cells retains their cell to cell interactions as well as collective polarity. Wound healing assay or scratch assay is a 2D *in vitro* technique used to study collective cell migration. In this assay, a scratch is made on the confluent cell monolayer resulting in the formation of a gap or wound which is monitored by taking pictures of the migrating cells at regular intervals of time. These pictures are then used to measure the speed of wound closure and thus quantify migration. Live-cell imaging using Time-lapse microscopy can be used for a more detailed study of cell migration behavior [20]. In order to reduce the effect of cell proliferation on gap filling, the readings are taken for a time period of 24 hours but this may vary depending on the cell line.

Another assay involving the response of single cells to various chemo-attractants is the transwell assay or the Boyden Chamber assay. This assay can be used for both adherent and non-adherent cells wherein the cells are placed in a serum-free medium on one side of a porous membrane and analyzed on the basis of the cell's ability to migrate through the pores to the other side. Cell migration can be quantified by counting the cells that have traversed through the membrane towards the higher concentration of chemoattractant [21]. A drawback of this assay is visualizing the cells and their morphology while migrating through pores due to the transitive state of cells [22].

2.3 *In vitro* tumorigenicity assay

Cancer is one of the most frightful diseases in both developing and developed countries and imparts a major health burden to the society. Tumorigenicity is the tendency of the cultured cells to form tumors. The two common *in vitro* tumorigenicity assays are - Colony-forming assay and Tumorsphere assay. The colony-forming assay is also referred to as clonogenic assay that analyzes the potentiality of a single cell to undergo a clonal expansion to form a colony composed of a minimum of 50 cells [23]. This assay is usually used to distinguish there productive viability of untreated cells from the cells that are treated with ionizing radiation or cytotoxic agents. It is also used to study the stemness and the clonogenicity of stem cells [24].

Colony forming assay is performed using the soft agar method. The basic steps involved in this assay are treating the cell monolayer in the flask, seeding the required number of cells on the agar and incubate for 1–3 weeks, fixing and staining the colonies and finally observing the colonies under the stereomicroscope [23]. Another *in vitro* tumorigenicity assay is tumorsphere assay which analyzes the potential of cancer stem cells (CSCs), a major cause of tumor initiation, progression, and recurrence after treatment. This assay is carried out under non-adherent conditions and serum-free medium supplied with growth factors of choice leading to the proliferation of CSCs and formation of spheres whereas the non-CSCs undergo apoptosis due to loss of adherence and abundant nutrients [25].

2.4 Cell invasion assay

Cell migration is an important process in biology where the cells changes and reaches their destination within a proper environment, in order to execute their respective function. It is a normal physiological process that takes place in nearly all forms of organisms. However, changes or deregulation of any kind in the pattern of cell migration or invasion are an indication of pathological conditions including inflammatory diseases and cancer metastasis, with the latter being the most explored one [21]. There are various biological methods that are commonly employed in the scientific community to study the above-mentioned events in depth namely, the cell culture wound-healing assay, the transwell migration, and invasion assay, individual cell-tracking assay, and spreading assay. These assays aim to provide relevant information pertaining to the pattern of cell migration or its response to chemoattractant(s).

2.4.1 The cell culture wound-closure assay

It is the simplest of all methods in determining the migration of whole-cell masses altogether. Going further in detail, it can be used to interpret individual cell's morphological characteristics and phenotypes during migration. Measuring the closed distance compared to the control over regular intervals of time shows specific migration changes or phenotype that was unknown in the past [26].

2.4.2 The transwell migration and invasion assay

The transwell migration and invasion assay are used to determine the capability of single cells to respond to various chemoattractant(s) including chemokines, growth factors, lipids, or nucleotides. It also contributes to assessing differential cell migration due to the over-expression of a receptor. It also identifies and characterizes the key regulators participating in cell migration [26].

2.4.3 Individual cell-tracking assay or single-cell tracking assay

Conducting single-cell tracking and its live imaging under appropriate conditions adds to the overall advantage of cell migration assay. The software includes a time-lapse video-microscopy protocol comprising of post-processing tracks of the cell populations with single-cell resolution. It greatly helps to understand the cell biology and lineage progression of distinct cell populations [27].

2.4.4 Cell spreading assay

In this type of assay, the spreading process of individual cells is seen and recorded with the help of Differential Interference Contrast microscopy (DIC). The spreading state is recorded every 5 seconds with a Charge-Coupled Device (CCD) of the camera, producing high-quality grayscale images. The process of taking images could extend to several hours [28].

2.5 Hybridoma technology and monoclonal antibodies

Antibodies, one of the major elements of the immune system are the glycoproteins produced by the immunoglobulins; B-cells provide protection against invading pathogens. The antibodies are highly specific and selective, thus have been used as an extraordinary tool in bioengineering and biomedical research for many years. The antibodies are majorly classified into two categories, Monoclonal Antibodies (mAbs) and Polyclonal Antibodies (pAbs) are based on their origin from the lymphocytes. mAbs are produced by only B lymphocyte or B cells and are monospecific. Due to this property, they possess high specificity and affinity towards a single epitope of an antigen whereas pAbs are produced by different B-cells and possess different affinities for multiple epitopes of a specific antigen. Since mAbs are highly specific, they are produced on a large scale through culturing of antibodies-producing cells widely known as 'Hybridomas', which are commonly derived from mice, and the method is known as 'Hybridoma Technology [29].

Hybridoma technology was discovered and developed by two eminent scientists, Georges Kohler and Cesar Milstein in 1975 and is considered to be one of the biggest breakthroughs. It has proved to be a robust, effective, and successful methodology employed in the field of biotechnology and biomedical research that solely deals with mAb isolation. The B cells go through the antibody maturation process in the germinal centers of secondary lymphoid tissues (for example, lymph nodes, spleen, tonsils, and Peyer's patches). Upon proliferation, certain mutations are experienced by the B cells, specifically in the genes encoding the variable region of the antibodies that helps in the selection for high-affinity tight binding to the corresponding antigen. The overall resulting antibodies by B cells consist of a natural pairing of the light chain and variable heavy chain genes with constant region genes. This region contains Class Switch Recombination (CSR) differentiates from the hybridoma technology in which CSR is absent [29].

Following are the steps employed for the production of monoclonal antibody by hybridoma technology.

2.5.1 Isolation of antibody-producing B lymphocyte

The mouse/mice is/are immunized every 2–3 weeks with red blood cells taken from sheep in order to produce the B cells. These antibodies are isolated from the spleen cells of mice.

2.5.2 Screening of mouse for production of antibody

After the process of immunization, the blood samples are taken from the mouse to determine the serum antibody titer. When the titer reaches the optimal level, the mouse is boosted by injecting antigen 3 days prior to fusion with myeloma cells [30].

2.5.3 Fusion of B cells with myeloma cells

Fusion of isolated spleen cells (limited life span) with tumor lymphocytes (immortal) with the help of PEG (Polyethylene Glycol) leads to the development of hybridomas with an unlimited life span.

2.5.4 Culturing of hybridomas

Hybridomas are grown in a selective medium containing Hypoxanthine, Aminopterin and Thymidine (HAT). Aminopterin present in the media blocks pathway for nucleotide synthesis, making the cells dependent on the alternative pathway which is not evident in myeloma cells.

2.5.5 Screening and selection of the desired colony

The cells are screened and chosen or selected for production of antibodies with the desired specificity.

2.5.6 Culturing of the selected hybridoma cells on large scale

The cells are cultured and used for the production of large quantities of antibodies [31].

2.5.7 Storage for future use

The cells are frozen and stored for future use in therapeutics.

2.6 Gene therapy

Gene, the fundamental biological unit of heredity that constitutes an ordered sequence of nucleotides present in chromosomes. The functional aspect of a gene is to encode a protein or RNA molecule inherited from parents such as texture and color of the hair and eyes. Any kind of alterations/mutations in a gene sequence can lead to abnormal functionality of the genes. Gene therapy is a modern type of experimental technique in the medical field which involves rectifying the non-functional or malfunctioning of genes by replacing them with healthy and functional genes. Several approaches have been implemented by researchers in terms of correcting a mutated gene with a healthy copy of the gene or by inactivating the mutated gene causing disease. It has been widely studied for various diseases such as immune deficiency, blood disorders, eye problems, metabolic disorders, regeneration of nerve cells, and cancer [32]. The first case of gene therapy was discovered in the 1990s whereby a functional Adenosine Deaminase (ADA) gene was incorporated in the white blood cells of the patient, replacing the non-functional ADA [33]. This application led to interesting results with the immune systems and hence, was considered the most reliable technique.

There are two main methods for gene therapy such as- *Ex-vivo* gene therapy *and In-vivo* gene therapy. The former is the transfer of genes into patient cells outside the body and the latter one is the transfer of genes directly to cells inside the body. To carry this, several techniques are used like- direct or liposome-mediated injection of DNA, calcium phosphate transfection, electroporation, dendrimers, hybrid methods, retrovirus, and other viral vectors. Clinical conditions on which gene therapy has been applied are as follows:

2.6.1 Parkinson's disease (PD)

The strategy of gene therapy has been applied to this disease in order to improve the advanced symptoms of PD. Gene therapy was applied to transfer 'Glutamic Acid Decarboxylase (GAD), a chemical produced by a gene into the basal ganglia. GAD showed an increased amount of a neurotransmitter called as Gamma-Aminobutyric Acid (GABA), responsible for inhibiting brain signals and decreasing activity in the nervous system Decreased GABA activity leads to certain brain-related disorders [34].

2.6.2 Alzheimer's disease (AD)

AD and other frontotemporal dementias (FTDs) are caused by the accumulation of amyloid-β peptide (Aβ) and protein tau in the brain. It is characterized by having memory loss, difficulty in learning and communicating along with the inability to organize things. The use of recombinant Adeno-Associated Viruses (rAAVs) has provided new ways for studying AD and other related neurological disorders [35]. Such strategies or approaches have added novel dimensions to medical treatments.

2.6.3 Cystic fibrosis

Cystic fibrosis is a disease known to affect the lungs primarily. Its symptoms include inflammation, airway obstruction leading to respiratory tract infection and deformity.

Insertion of the Cystic Fibrosis Transmembrane Regulator (CFTR) gene directly into the epithelium cells of the respiratory tract bear the capability to lessen the symptoms but not totally cure the disease in patients suffering from cystic fibrosis [36].

2.7 Cell therapy

Cell-based therapy is one of the most important and well-known forms of all treatments in the fields of modern science & medicine. It is not only a curative option for treating deadly or threatening diseases but is also making 'Regenerative Medicine' the most vital technique in health care with the specific goal of replacing diseased cells, tissues or organs and thereby restoring their normal function(s) [4]. Over the years, there has been a gush of interests and work done in understanding the potential of stem cells. They are the cells found naturally in the living bodies, characterized by two defining properties of eternal self-renewal and the propensity to differentiate into an adult cell type. There are three main types of stem cells: Totipotent (a cell developing into a healthy organism independent of the permissive environment), Pluripotent (a cell developing into any type of adult cell) and Multipotent (a cell developing into a limited type of cell) [37].

Following is the account of different stem cells used for the treatment of various diseases:-

2.7.1 Pluripotent stem cells

Reportedly, pluripotent cells have been used successfully to treat animals per se. Animals diagnosed with diabetes are incorporated with cells containing insulin responsive to glucose levels. Additionally, the treatment of the animals suffering from acute spinal cord injury and visual impairment is performed with myelinated neurons and retinal epithelial cells, respectively. Researchers are still conducting studies with the use of pluripotent stem cells to cure several disorders such as Parkinson's disease, muscular dystrophy and heart failure.

2.7.2 Induced pluripotent stem cells

The stem cells created artificially from normal adult somatic cells through co-expression of genes and factors are known as Induced Pluripotent Stem Cells (iPSCs). These are important for maintaining the characteristic properties of Embryonic Stem (ES) cells. Some reports have stated the successful use of iPSCs in conditions like Parkinson's disease, spinal muscular atrophy, cardiac diseases, blood disorders, diabetes, amyotrophic lateral sclerosis, Huntington's disease, and familial dysautonomia.

2.7.3 Multipotent stem cells

The multipotent stem cells derived from bone marrow (Hematopoietic stem cells) have been used in the 1960s to treat cancer conditions like leukemia, myeloma and lymphoma. Mesenchymal stem cells with the capability of forming whole joints in mouse models have been used regenerating bone and cartilages form. Curing heart ailments are still under clinical trial.

3. Applications of 3D cultures *in vitro*

Spurred by the recent advent in cell culture technologies, three-dimensional (3D) cell culture is paving the way in promoting tissue organization and cell differentiation by triggering tissue-based diseased microenvironment. An ideal 3D cell culture system generally composed of tightly bound tissues that involve cell–cell fluent interaction almost mimicking the extracellular matrix (ECM) that is highly dynamic and includes scaffolds of cells in a fluid that enhances them to differentiate (**Table 1**). The key parameter of a 3D culture environment is the ability to organize the spatial arrangement of cells with other surrounding cells along with physical constraints [8]. This significant approach has gardened great focus on understanding complex cellular biology and their responses by validating mammalian tissue studies via linking the gap between *in vitro* and *in vivo* environments. The two-dimensional (2D) cell cultures lack several features that 3D cultures impart such as tissue-specific architecture and complex cellular interactions that make them poor models for complex diseases. Based on the process of preparation, 3D techniques are categorized into (i) scaffold-based, (ii) scaffold-free culture systems. Scaffold-based technique efficiently is more responsive towards cell-to-ECM connection because of their potentiality of mimicking ECM whilst scaffold-free technique persuades physiological and cellular gradients. Scaffold or matrix is designed according to the tissue of interest, higher is the complexity of the scaffold, and the higher is the difficulty to extract the cells for analysis [9]. Majorly, the

Approach	Merits	Demerits
3D Spheroids	• High reproducibility	• Expensive to prepare
	• Therapeutic inhibition to drug exposure can be easily evaluated through image analysis	• Optimization protocols for each cell line is required
	• Constant perfusion	• Difficult light matter interactions for large spheroids
	• Lower consumption of reagents	• Differences in spheroids' diameters,
	• Control of shear stress and pressure on cells	• Low-throughput
	• Capable of imbibing large amount of water or biological fluid	• Labour intensity
Hydrogels	• Ease of maintenance	• Low mechanical strength
	• Amenable of controlling the micro-tissue size and large amount of micro-tissues per plate	• Difficult to handle
		• Expensive
Organoids	• Amenable to high-throughput screening	• Absence of microenvironment
	• Long lived organoid production from single cells	• Optimization protocols are not globally standardized.
		• Organoid cultures rapidly die due to contamination
Cancer co-culture Models	• Easily evaluate cell–cell interactions of cancer microenvironment	• Microbial contamination
	• Provide fluid flow	• Static condition
	• Easy to handle and quantify	
	• Relevant mechanical cues	
Organ-on-a-chip	• Enable stable co-culture of living human cells	• Architectural complexity of developing human tissues and organs
	• Good control over microenvironment	• Difficulty in standardization and scale-up

Table 1.
Merits and demerits of different 3D cell culture techniques.

scaffolds manufactured are polymeric hard material-based support, microfluidic-based assembly, hydrophilic glass fiber, and organoids. In contrast, scaffold-free systems form cellular aggregates called spheroids that have evolved with improvement in their techniques such as magnetic levitation, liquid overlay (low adhesion plates), hanging drop microplates, and spheroid microplates yet the demand for 3D scaffolds preferentially increased due to their immense property of modulating the behavior of cultured cells according to the matrix in which they are cultured.

3.1 3D Spheroids

3Dspheroids or multi-cellular aggregates are spherical micro-sized cellular constructs that are produced from numerous gamuts of cell types, originally from scaffold-free systems. The most characteristic features of 3D spheroids are the ability to recapitulate a cell's typical physiological behavior, cellular heterogeneity, gene expression, cell–cell signaling, and structural architecture with respect to cell–cell contact [10]. Various types of 3Dspheroids include embryonic bodies, tumor spheroids (spheres of different tumorcells), hepatospheres (spheres of hepatic cells), neurospheres (spheres of different cell types of the central nervous system (CNS)) and mammospheres (spheres of mammary glands) [38]. An ideal 3D spheroid constitute ECM components such as proteoglycans, laminin, collagen, fibronectin, tenascin, and glycosaminoglycans [39] which tightens the spheroid density with close ECM-cell and cell–cell anchors eventually increase interstitial fluid pressure (IFP). Depending on the primary amount of cells seeded, the size of spheroid increases with an elevation in cell number, oxygen, and nutrient gradients equivalent to the tissue of interest [5]. Alongside, the different techniques enabling spheroid cultures are illustrated further.

Hanging drop technique is a non-scaffold method wherein a drop of media containing cells are suspended inversely on the lid of the culture dish (bottom-less and open) such that there is no surface provided for the cells and tend to hang. This attempt forms a cluster called spheroid at the tip of the droplet when cultured for a longer period [40]. Spheroids formed through hanging drop cultures have fetched considerable stance in cell culture technology with 100% reproducibility owing to ubiquitous applications in cancer research [41], toxicity testing in hepatocytes [42], and constructing cardiac spheroids [43]. Another method involves the use of a liquid overlay that eases the formation of aggregates and commercially produced as low adhesion plates. These spheroid microplates contain either hydrophilic or hydrophobic coating with V-shaped bottom and allow mild attachment to the surface such that the cells tend to self-aggregate and form spheroid. Unlike the hanging drop technique, low adhesion plates generate one spheroid per plate that signifies its importance for multicellular culture. This ensures a medium-throughput screening that requires no modification in spheroid formation [44]. Spheroids can also be cultured with the use of magnetic nanoparticles with the application of the magnetic field. The process is called magnetic cell levitation that is highly applied to produce spheroids of mesenchymal stem cells and tissue engineering [45, 46]. An *in vivo* study showed that human glioblastoma cells levitated by a magnetic approach closely mimicked the protein expression of human glioblastoma tumor xenografts [47].

3.2 Organoids

Organoids refer to the primary cultures derived from cell aggregates through *in vitro* process that is grown in 3D gels containing ECM to produce organ-like buds with the application of either physical support (cell adherence) or biochemical cues (signaling pathway modulation). Various types of cells such as embryonic, adult, primary, and stem cells are utilized for the development of organoids. Based on organ-like structures formed, organoids are classified into tissue and stem cell organoids. The application of organoids has helped in producing numerous *in vitro* organoids such as rectal [48], gastric [49], lung [50], liver [51], pancreas [52], retina [53], thyroid [54], kidney [55] and intestine [56] that had successfully recapitulated the structural and functional motif of real organs. 3D organoids are extensively used by researchers to decipher the toxicity analysis, examine the genetic pathologies and investigate the local immune responses to infections. In addition to this, current reports have suggested the promiscuous application of organoids in platforms like transcriptomics and proteomics technologies. One such example illustrates the interaction study between Zika virus and Toll-like receptor 3 is performed by the generation of cerebral organoids from embryonic stem cells [57]. They have been also used as models for distinct genetic diseases. For instance, a study applied the rectal organoid model of cystic fibrosis for the investigation of the potency of transmembrane regulator-modulating compounds [48]. Besides, the tubular organoids model of polycystic kidney disease was also used to unravel the cause for cyst formation [55]. Apart from this, organoids have been an excellent source of models to understand the depth of neurodegenerative diseases viz.; Alzheimer, Parkinson's, HIV, diabetes, or cancer.

3.3 Cancer co-culture models

Cancer cell lines have emerged as an eminent tool for comprehending complex physiology of cancers. The cell cultures have eased the outlook in preclinical research to understand the process of disease, morphological changes occurring in tissue, gene function, cell biology and tissue engineering [58]. They have evolved with immense features of offering homogenous samples without any sort of modification and variations. However, a big leap was noted when monolayer cell cultures (2D) obtained from solid tumors were incapable of mimicking the structural elements of tumor microenvironment. Thusly, 3D cancer cell culture models have placed an enduring platform recently whereby ECM in 3D construct is same as that of original cell culture and imparted knowledge of predicting tumor response to treatment [59]. The application of 3D cell culture models of tumors have ought to manifest typical properties of tumor microenvironment such as gene and protein expressions, morphology, angiogenesis, malignancy and invasiveness. From this standpoint, 3D tumor cell culture models scintillate anticancer therapeutics and cancer drug discovery. To date, a vast content of literature owes the significance of these 3D co-cultures models in varying applications. In a study, tumor-associated macrophages (TAM) or cancer-associated fibroblasts (CAF) and gelatin hydrogel microspheres (GM) have been applied to produce cancer co-culture models from different cancer cells including HepG2 (liver), MCF-7 (breast) and WA-hT (lung) in order to inspect sustained release of drugs. They induced metastatic proteins involved in epithelial-mesenchymal transition (EMT) with transforming growth factor-β1 (TGF-β1) and reported elevation in N-cadherin and Vimentin proteins with deceleration in E-cadherin protein [58]. Recently, cancer co-culture models evinced interest in numerous approaches such as 3D breast cancer co-culture models obtained from MCF-7, MRC-5 and MDA-MB-231 tumor cells were used in investigating radiation-induced fibrosis [59], tumor-associated fibroblast differentiation [60] and development of immunotherapies [61], 3D lung cancer co-culture models derived lung squamous carcinoma and Non-Small Cell Lung Cancer Cells (NSCLC)fromTUM622, A549 and Colo699 tumor cells were utilized to explore tumor-stroma interactions [62, 63], 3D renal cancer co-culture models formed from

Caki 1 (skin metastasis derived) and ACHN (pleural effusion derived) were sought for determining the efficacy of produced 3D models in stem cell physiology research and drug toxicity screening [64]. 3D colon cancer co-culture models acquired from LS 174 T, HCT 116, Colo205, MCF7, SW480, SW620, CCD-18Co, Caco-2, HT-29, and H446 have also been used to explore tumor-stroma interactions [65].

3.4 Tissue co-culture models

In vitro tissue models with the use of co-culture cells have emphasized greater applications to represent varying mechanisms of human body which is a daunting task. These models have served a vital role over several animal models that once were used to examine human physiology and pathophysiology. The major limitation of animal models was the failure of mimicking true human facets and their ethical constraints. This led to fetch insights into development of tissue models as a research tool from co-cultures such that the created models would possibly recapitulate the natural microenvironment of cells and examine the pathophysiological bases of diseases. Distinct *in vitro* 3D tissue models have been achieved with the approach of tissue engineering comprising human characteristics with increased complexity as compared to the 2D monolayers. Some of the examples of 3D models constructed from tissue co-cultures include kidney [66], neuro-glia [67], lung [68], liver [69], ovary [70] and intestine [71] that have potentially predicted and represented physiological responses of the original culture of cells. Most probably, primary cells are used as they possess feasible *in vivo* features of not being immortal, incapable of getting transformed and consist of limited survival time in culture. Pertaining to these advantages, they have been applied to develop models that would combat various disease and physiological studies. Reportedly, tissue models have been revolutionized in terms of investigating multiple changes in real-time processes. A pulmonary endothelium model was constructed to investigate massive inflammation in patients with acute respiratory disease syndrome (ARDS). The authors performed this study using lavage samples of the patients for determining the etiology of ARDS that took place during the process of disease [72]. In addition, endothelial cells have also played a key role in constructing 3D tissue models with their fascinating physiological roles. For instance, *in vitro* gastrointestinal epithelial cell cultures derived from adult murine colon allowed the authors to analyze epithelial cell–cell interactions, microbiological infections and cellular signaling [73]. Another instance showed the potential of hepatocyte tissue cultures in maintaining the cancer cell hierarchy in human hepatocellular carcinoma [74].

3.5 Organ-on-a-chip

Organ-on-a-chip is a biomimetic system that uses fabrication of computerized microchips and microfluids consisting of living cells, mimicking the natural environment of organs from which it is been created. There are several factors that made organ chips be listed in "Top Ten Emerging Technologies" in the World Economic Forum [75] such as shear force, tissue-boundaries, concentration gradients, tissue–organ interactions and cell patterning. Organ chips have intensified in the field of drug therapeutics for their ability of high throughput screening. **Table 2** summarizes the recent researches carried out using various organ chips. These organ chips use microtechnology that provides nutrients to the cells for their better growth and proliferation. Microfluids are one such component that has been used in various studies for efficient treatment in drug sensitivity testing [86]. Talking of this notion, a microfluidic chip was produced in order to monitor and document real-time impedimetric biosensor changes. Other organ-on-a-chip models such as blood–brain barrier chips

Organ type	Incorporated cell types	Organ-specific properties	Ref.
Lung	Primary lung alveolar epithelial cells	• Breathing movements	[76, 77]
	Primary lung endothelial cells	• In vivo functionality array of tiny alveoli	
		• Human lung parenchyma (lung alveoli and ultrathin air–blood barrier)	
		• Recreates the native viscoelastic microenvironment of the cells	
	Human vascular endothelial cells	• Alveolar capillary barrier in the human lung	[78]
	Human alveolar epithelial cells		
Skin	Peripheral perfusion fluid (PPF)	• Franz diffusion cell system	[79]
		• Drug absorption across the dermal barrier	
		• Microfluidic Diffusion Chamber (MDC)	
Liver	Hepatic cell lines	• Hepatoprotective effect assessment	[80]
		• Hepatic activity (cell viability, albumin synthesis, urea secretion, and cyto-chrome P450 enzyme activities)	
		• Drug Screening and Toxicity Testing	
Kidney	Human podocytes Glomerular endothelial cells	• Human glomerular filtration barrier	[81]
		• Functions and structure of the glomerulus	
	Human-derived renal proximal tubule epithelial cells	• Cell Polarization-Dependent Cisplatin Toxicity	[82]
		• Proximal tubule	
Heart	Cardiomyocytes	• Contractile behavior (contraction force, frequency, and synchronization) of a 3D cardiac tissue construct • Three-dimensional beating tissue from human cardiomyocytes	[83, 84]
Pancreas	Human pancreatic beta-cell line	• Human pseudo-islets in biomechanical flow conditions	[85]

Table 2.
Summary of recent organ-on-a-chip models.

have been developed to represent the *in vivo* architecture of brain involving micro blood vessels by using type 1 collagen hydrogel. Another model entails the significance of human-on-a-chip that depicts the normal human physiology in combination with single organ chip within a microfluidic system that ultimately forms multi-organ chip [87]. One study dealt with an *in vitro* 3D-tumor-on-a-chip device that illustrated its importance in quorum sensing phenomenon in tumor cells activated by salmonella [88]. Hence, organ-on-a-chip has been diversified in many scientific platforms due to their efficient physiological bio-mimicry.

3.6 Patient-derived cells

Animal models used in laboratories have been greatly avoided due to the fact that they are costly and require a large number of laborers. This approach was replaced by the use of *in vitro* models wherein despite having several advantages; still, the application is constrained due to poor cell-to-cell and cell-to-matrix interactions. *In vitro* culture of cells has been observed to acquire multiple genetic and epigenetic variations that eventually make the cells lose their originality. The above-listed models comprise their own merits and demerits in respect of cellular response, cellular composition, and structural features. The above-mentioned models consist of a few advantages and disadvantages with respect to cellular composition, mimicking the *in vivo* physiology of original tumor architecture, tumor microenvironment (TME) and the response to different exogenous stimuli. Therefore, patient-derived cells have come into the picture which is generally a co-culture-based technology that is grown in a culture medium supplemented with all sorts of nutrients [89]. Different studies have exemplified the use of patient-derived cells as a preclinical model in drug discovery (screening and responses) in several types of cells. Patient-derived xenograft (PDX)models are made of minute pieces of tumor tissue of the surgical patients, utilized for implantation into an immune-deficient mouse. In a study, Fong et al. used PDX models from prostate cancer to investigate tumor-stromal interactions via the use of a 3D hydrogel system [90]. Likewise, in another study, Liu et al. demonstrated the establishment of patient-derived cell cultures from colorectal cancer cells of biopsies of cancerous and non-cancerous tissue that could grow in *in vitro* culture indefinitely by recapitulating exactly the same phenotypic and genotypic features of the original tissue [91]. Some of the researchers have also elaborated patient-derived 3D culture using a scaffold-based organoid culture that is prepared to preserve the genomic features of the original tissue [92].

4. Three-dimensional cell culture in drug discovery and development

Drug discovery is a lengthy and time-consuming process that undergoes several stages of testing and optimization. This encompasses identification of the target, lead discovery, pre-clinical validation, and clinical trials [3]. Due to the constant failure of drugs in Phase II and Phase III clinical trials, there has been constant pressure on the pharmaceutical industry to seek more novel drugs with lower side effects and cost-effectiveness. 3D cell culture has emerged as a significant high-throughput system that has uplifted the standards of cell culture [93]. Specifically, spheroids are considered the most reliable model for testing drugs in various diseases because of their capability of resembling the natural environment of original tissue [93]. The spatial organization of spheroids in different layers of cells leads to cellular death by forming reactive oxygen species [94]. In the case of investigating the effect in 3D spheroids, fluorescence microscopy plays a key role in determining pharmaceutical dispersion within spheroids (eg-doxorubicin and epirubicin) [95]. The capital importance of any drug testing involves cell-based assays that are efficient enough and easily reproducible compared to expensive animal models. Cell-based assays have shaped the physiological relevance of 2D cultures [96]. While the reaction may vary from technique to technique such as cell viability, proliferation, signaling and migration and drug to drug for achieving better sensitivity. It is now broadly accepted that compared to 2D cultures, 3D models serve the resemblance of the natural environment of original tissue efficiently and differently in 3D environments. Research has nested stance on novel 3D culture technologies that impart functional basis of tissues such as spheroids and organoids [97]. A study used 3D hydrogel-based model for the determination of drug sensitivity in HepG2 cell lines by comparing cytotoxicity effect with cytotoxicity (CT_{50}) and lethal dose (LD_{50}) values [98]. Organoid 3D models also aid as a resourceful tool for modeling neurodevelopmental disorders [98]. Microfluidic chips have also been utilized in drug sensitivity testing whereby a study elaborated its efficacy in lung cancer which was in combination with stromal cell lines [98]. Evaluation of absorption, distribution, metabolism, excretion, and toxicity (ADMET) of the drug is primarily examined in *in vitro* cell culture experiments. 3D cell cultures have fostered drug pharmacokinetics in several studies with the implementation of various types of cell culture models.

5. Cell based manufacturing of therapeutic proteins

Therapeutic proteins production using human cell lines has greatly influenced different medical areas including biopharmaceutical research and vaccine production. Mammalian cell lines prove futile in protein production due to their likelihood of possessing post-translational modifications (PTMs) achieved from recombinant proteins that are in accordance with the endogenous human proteins. These cell lines show exquisite specificity to produce similar proteins to those in humans naturally synthesized, an advantage over mammalian expression systems [99]. One of the most routinely and high yields of proteins production is performed by using cell-based expression systems such as Chinese hamster ovary (CHO) a cell line that constitutes major advances such as accomplishment of gene amplification, specific productivity, better selection strategies, and devising greater expression units and advanced hosts. CHO cells have established their safety profile for 20 years from the production of its first recombinant biotherapeutic protein in 1986 [100]. Other human cell lines such as BHK-21 cells are used for the generation of few coagulation factors such as factor VIII [101]. There are two vital human cell lines namely, HEK293 and HT-1080 that are used to manufacture licensed products of human PTMs. The advancement in protein-based drug development and technologies has driven more towards the therapeutic proteins market that comprises of sales of these therapeutic proteins. The methods that are involved in the production of these proteins are pegylation, glycoengineering, albumin fusion, Fc-fusion, product purity, targeting, and functionality of therapeutic protein drugs. Few examples of therapeutic protein drugs which has been produced using protein engineering technologies and approved by the Food and Drug Administration (FDA) from the past five years are imiglucerase, Belimumab, alfa, coagulation factor IX recombinant human and albiglutide [102]. A French pharmaceutical company named Sanofi accomplished a great achievement of strengthening its R&D strategy with the best proprietary therapeutic proteins production pharmaceutical company, Ablynx, for a nanobody technology platform.

6. Conclusion

In particular, a plethora of research studies have shed light on the fact that in spite of the availability of advanced organ-on-chip technologies and bioengineered 3D models, the application is limited by drug companies due to their relatively novel approach which is more likely requires to undergo further validation and characterization. Moreover, 3D cell culture models with high-throughput screening in combination with high-content leads to the identification of clinically relevant compounds. However, still many difficulties are being faced as 3D cell cultures do not meet certain criteria in the drug discovery process with regard to size, morphology, complexity, and protocol for assaying. It requires ample standardization and optimization to extract successful specific phenotypes for drug screening. Thus, there are few 3D models that are constrained for their restricted access due to limited permeability. Following the advances in protein therapeutics, more improvements in generating sophisticated therapeutic protein products will be developed for better futuristic research.

Acknowledgements

Authors thanks, Dr. D. Y. Patil Biotechnology and Bioinformatics Institute, Dr. D. Y. Patil Vidyapeeth, Tathawade, Pune, for providing necessary facilities. This work was supported by Dr. D. Y. Patil Vidyapeeth seed grant (DPU/14/2016, dated 06/01/2016).

Conflict of interest

There are no conflicts of interest.

Abbreviations

2D	Two-dimensional
3D	Three-dimensional culture
Aβ	Amyloid-β peptide
ADMET	Absorption, distribution, metabolism, excretion and toxicity
ADA	Adenosine Deaminase
AD	Alzheimer's disease
ARDS	Acute respiratory disease syndrome
CAF	Cancer-associated fibroblasts
CCD	Charge-Coupled Device
CFTR	Cystic fibrosis transmembrane regulator
CHO	Chinese hamster ovary
CNS	Central nervous system
CSCs	Cancer stem cells
CT50	Cytotoxicity 50 percent
CSR	Class Switch Recombination
DIC	Differential Interference Contrast microscopy
ECM	Extracellular matrix
ES	Embryonic Stem
EMT	Epithelial-mesenchymal transition
FTDs	Frontotemporal dementias
GABA	Gamma-Aminobutyric Acid
GAD	Glutamic Acid Decarboxylase
GM	Gelatin hydrogel microspheres
HAT	Hypoxanthine, Aminopterin, Thymidine
IFP	Interstitial fluid pressure
iPSCs	Induced pluripotent stem cells
LDH	Lactate dehydrogenase
LD50	Lethal dose 50 percent
MTT	3-(4,5-dimethythiazol2-yl)-2,5-diphenyl tetrazolium bromide
MTS	3-(4,5-dimethylthiazol-2-yl)-5-(3-carboxymethoxyphenyl)-2-(4-sulfophenyl)-2H-tetrazolium

MDC	Microfluidic Diffusion Chamber
mAbs	Monoclonal Antibodies
NAD	Nicotinamide adenine dinucleotide
NADH	Reduced nicotinamide adenine dinucleotide
NADPH	Nicotinamide adenine dinucleotide phosphate
NSCLC	Non-Small Cell Lung Cancer Cells
pAbs	Polyclonal Antibodies
PDX	Patient-derived xenograft
PD	Parkinson's Disease
PPF	Peripheral perfusion fluid
PTMs	Post-translational modifications
rAAVs	Adeno-Associated Viruses
SRB	Sulforhodamine B
TAM	Tumor-associated macrophages
TGF-β1	Transforming growth factor-β1
TME	Tumor microenvironment
XTT	2,3-Bis-(2-Methoxy-4-Nitro-5-Sulfophenyl)-2H-Tetrazolium-5-Carboxanilide.

Author details

Sangeeta Ballav[1], Ankita Jaywant Deshmukh[2], Shafina Siddiqui[2], Jyotirmoi Aich[2] and Soumya Basu[1]*

1 Cancer and Translational Research Laboratory, Dr. D.Y. Patil Biotechnology and Bioinformatics Research Institute, Dr. D.Y. Patil Vidyapeeth, Pune, Maharashtra, India

2 School of Biotechnology and Bioinformatics, D.Y. Patil Deemed to be University, Navi Mumbai, Maharashtra, India

*Address all correspondence to: soumya.basu@dpu.edu.in

References

[1] Hudu SA, Alshrari AS, Syahida A, Sekawi Z. Cell Culture, Technology: Enhancing the Culture of Diagnosing Human Diseases. J ClinDiagn Res. 2016;10(3):DE01-DDE5. DOI:10.7860/JCDR/2016/15837.7460.

[2] Verma A, Verma M, Singh A. Animal tissue culture principles and applications. Animal Biotechnology. 2020;269-293. DOI:10.1016/B978-0-12-811710-1.00012-4.

[3] Smith SC, Baras AS, Lee JK, Theodorescu D. The COXEN principle: translating signatures of *in vitro* chemosensitivity into tools for clinical outcome prediction and drug discovery in cancer. Cancer Res. 2010; 70, 1753-1758. DOI: 10.1158/0008-5472.CAN-09-3562.

[4] Michelini E, Cevenini L, Mezzanotte L, Coppa A, Roda A. Cell-based assays: fuelling drug discovery. Anal Bioanal Chem. 2010; 398(1):227-238. DOI: 10.1007/s00216-010-3933-z.

[5] Kapałczyńska M, Kolenda T, Przybyła W, Zajączkowska M, Teresiak A, Filas V, Ibbs M, Bliźniak R, Łuczewski Ł, Lamperska K. 2D and 3D cell cultures - a comparison of different types of cancer cell cultures. Arch Med Sci. 2018;14(4):910-919. DOI:10.5114/aoms.2016.63743.

[6] Verma A, Verma M, Singh A. Animal tissue culture principles and applications. Animal Biotechnology. 2020; 269-293. DOI:10.1016/B978-0-12-811710-1.00012-4.

[7] Kamińska K, Szczylik C, Bielecka ZF, Bartnik E, Porta C, Lian F, Czarnecka AM. The role of the cell-cell interactions in cancer progression. Journal of cellular and molecular medicine. 2015; 19(2), 283-296. DOI: 10.1111/jcmm.12408.

[8] Ravi M, Paramesh V, Kaviya SR, Anuradha E, Solomon FDP. 3D cell culture systems: Advantages and applications. Journal of Cellular Physiology. 2015;230(1):16-26. DOI: 10.1002/jcp.24683.

[9] Fennema E, Rivron N, Rouwkema J, van Blitterswijk C, de Boer J. Spheroid culture as a tool for creating 3D complex tissues. Trends in Biotechnology. 2013;31(2):108-115. DOI: 10.1016/j.tibtech.2012.

[10] Zanoni M, Piccinini F, Arienti C, Zamagni A, Santi S, Polico R, Bevilacqua A, Tesei A. 3D tumor spheroid models for *in vitro* therapeutic screening: A systematic approach to enhance the biological relevance of data obtained. Sci. Rep. 2016;6:19103. DOI: 10.1038/srep19103.

[11] Duval K, Grover H, Han LH, Mou Y, Pegoraro AF, Fredberg J, Chen Z. Modeling Physiological Events in 2D vs. 3D Cell Culture. Physiology (Bethesda). 2017;32(4):266-277. DOI: 10.1152/physiol.00036.2016.

[12] Freimoser FM, Jakob CA, Aebi M, Tuor U. The MTT [3-(4,5-dimethylthiazol-2-yl)-2,5-diphenyltetrazolium bromide] assay is a fast and reliable method for colorimetric determination of fungal cell densities. Appl Environ Microbiol. 1999;65(8):3727-3729. DOI:10.1128/AEM.65.8.3727-3729.1999.

[13] Lomakina GY, Modestova YA, Ugarova NN. Bioluminescence assay for cell viability. Biochemistry (Mosc). 2015;80(6):701-713. DOI:10.1134/S0006297915060061.

[14] Kumar P, Nagarajan A, Uchil PD. Analysis of Cell Viability by the Lactate Dehydrogenase Assay. Cold Spring HarbProtoc. 2018; 2018(6). DOI:10.1101/pdb.prot095497.

[15] Khodabukus A, Prabhu N, Wang J, Bursac N. *In Vitro* Tissue-Engineered Skeletal Muscle Models for Studying Muscle Physiology and Disease. AdvHealthc Mater. 2018; 7(15):e1701498. doi:10.1002/adhm.201701498.

[16] DeStefano JG, Jamieson JJ, Linville RM, Searson PC. Benchmarking *in vitro* tissue-engineered blood-brain barrier models. Fluids Barriers CNS. 2018; 15(1):32. DOI:10.1186/s12987-018-0117-2.

[17] Shen TW, Puccini B, Temnyk K, Herting S, Cardinal KO. Tissue-engineered aneurysm models for *in vitro* assessment of neurovascular devices. Neuroradiology. 2019; 61(6):723-732. DOI:10.1007/s00234-019-02197-x.

[18] Jakubowska W, Chabaud S, Saba I, Galbraith T, Berthod F, Bolduc S. Prevascularized Tissue-Engineered Human Vaginal Mucosa: *In Vitro* Optimization and *In Vivo* Validation. Tissue Eng Part A. 2020; 26(13-14):811-822. DOI:10.1089/ten.tea.2020.0036.

[19] Winkler J, Abisoye-Ogunniyan A, Metcalf KJ,Zena W. Concepts of extracellular matrix remodelling in tumour progression and metastasis. Nat Commun. 2020; 11(1), 5120. DOI:10.1038/s41467-020-18794-x.

[20] Grada A, Otero-Vinas M, Prieto-Castrillo F, Obagi Z, Falanga V. Research Techniques Made Simple: Analysis of Collective Cell Migration Using the Wound Healing Assay. J Invest Dermatol. 2017; 137(2):e11-e16. DOI:10.1016/j.jid.2016.11.020.

[21] Pijuan J, Barceló C, Moreno DF, Maiques O, Sisó P, Marti RM, Macià A, Panosa A. *In vitro* Cell Migration, Invasion, and Adhesion Assays: From Cell Imaging to Data Analysis. Frontiers in cell and developmental biology, 2019; 7, 107. DOI: 10.3389/fcell.2019.00107.

[22] Gau DM, Roy P. Single Cell Migration Assay Using Human Breast Cancer MDA-MB-231 Cell Line. Bio Protoc. 2020; 10(8):e3586. DOI:10.21769/bioprotoc.3586.

[23] Franken NA, Rodermond HM, Stap J, Haveman J, van Bree C. Clonogenic assay of cells *in vitro*. Nat Protoc. 2006; 1(5):2315-2319. DOI:10.1038/nprot.2006.339.

[24] Rajendran V, Jain MV. *In Vitro* Tumorigenic Assay: Colony Forming Assay for Cancer Stem Cells. Methods Mol Biol. 2018; 1692:89-95. DOI:10.1007/978-1-4939-7401-6_8.

[25] Wang H, Paczulla AM, Konantz M, Lengerke C. *In Vitro* Tumorigenic Assay: The Tumor Spheres Assay. Methods Mol Biol. 2018; 1692:77-87. DOI:10.1007/978-1-4939-7401-6_7.

[26] Justus CR, Leffler N, Ruiz-Echevarria M, Yang LV. *In vitro* cell migration and invasion assays. J Vis Exp. 2014; (88):51046. DOI:10.3791/51046.

[27] Gómez-Villafuertes R, Paniagua-Herranz L, Gascon S, de Agustín-Durán D, Ferreras M O, Gil-Redondo JC, Queipo MJ, Menendez-Mendez A, Pérez-Sen R, Delicado EG, Gualix J, Costa MR, Schroeder T, Miras-Portugal MT, Ortega F. Live Imaging Followed by Single Cell Tracking to Monitor Cell Biology and the Lineage Progression of Multiple Neural Populations. J Vis Exp. 2017; (130):56291. DOI:10.3791/56291.

[28] Fardin MA, Rossier OM, Rangamani P, Avigan PD, Gauthier NC, Vonnegut W, Mathur A, Hone J, Iyengar R, Sheetz MP. Cell spreading as a hydrodynamic process. Soft Matter. 2010; 6(19), 4788. DOI:10.1039/c0sm00252f.

[29] Parray HA, Shukla S, Samal S, Shrivastava T, Ahmed S, Sharma C, Kumar R. Hybridoma technology a

versatile method for isolation of monoclonal antibodies, its applicability across species, limitations, advancement and future perspectives. IntImmunopharmacol. 2020; 85:106639. DOI:10.1016/j.intimp.2020.106639.

[30] Holzlöhner P, Hanack K. Generation of Murine Monoclonal Antibodies by Hybridoma Technology. J Vis Exp. 2017; (119):54832. DOI:10.3791/54832.

[31] Carvalho LS, Bravim da Silva O, Carneiro de Almeida G, Davies de Oliveira J, ParachinNS,Carmo TS. Production Processes for Monoclonal Antibodies. Intechopen; 2017. DOI:10.5772/61924(Chapter 10)

[32] Gonçalves GAR, Paiva RMA. Gene therapy: advances, challenges and perspectives. Einstein (Sao Paulo). 2017;15(3):369-375. DOI:10.1590/S1679-45082017RB4024.

[33] Aiuti A, Roncarolo MG, Naldini L. Gene therapy for ADA-SCID, the first marketing approval of an *ex vivo* gene therapy in Europe: paving the road for the next generation of advanced therapy medicinal products. EMBOMol Med. 2017; 9(6):737-740. DOI:10.15252/emmm.201707573.

[34] Axelsen TM, Woldbye DPD. Gene Therapy for Parkinson's Disease, An Update. J Parkinsons Dis. 2018; 8(2):195-215. DOI:10.3233/JPD-181331.

[35] Combs B, Kneynsberg A, Kanaan NM. Gene Therapy Models of Alzheimer's Disease and Other Dementias. Methods Mol Biol. 2016; 1382:339-366. DOI:10.1007/978-1-4939-3271-9_25.

[36] Marquez Loza LI, Cooney AL, Dong Q, Randak CO, Rivella S, Sinn PL, McCray PB. Increased CFTR expression and function from an optimized lentiviral vector for cystic fibrosis gene therapy. MolTher Methods Clin Dev. 2021;21:94-106. DOI:10.1016/j.omtm.2021.02.020.

[37] Aly RM. Current state of stem cell-based therapies: an overview. Stem Cell Investig. 2020;7:8. DOI:10.21037/sci-2020-001.

[38] Edmondson R, Broglie JJ, Adcock AF, Yang L. Three-dimensional cell culture systems and their applications in drug discovery and cell-based biosensors. Assay Drug DevTechnol. 2014;12(4):207-218. DOI:10.1089/adt.2014.573.

[39] Yeon SE, No da Y, Lee SH, Nam SW, Oh IH, Lee J, Kuh HJ. Application of concave microwells to pancreatic tumor spheroids enabling anticancer drug evaluation in a clinically relevant drug resistance model. PLoS One. 2013;8(9):e73345. DOI:10.1371/journal.pone.0073345.

[40] Tung YC, Hsiao AY, Allen SG, Torisawa YS, Ho M, Takayama S. High-throughput 3D spheroid culture and drug testing using a 384 hanging drop array. Analyst. 2011;136(3): 473-478. DOI: 10.1039/c0an00609b.

[41] Zhao L, Xiu J, Liu Y,Zhang T, Pan W, Zheng X, Zhang X. A 3D Printed Hanging Drop Dripper for Tumor Spheroids Analysis Without Recovery. Sci Rep 2019; 9, 19717. DOI: 10.1038/s41598-019-56241-0

[42] Hurrell T, Ellero AA, Masso ZF, Cromarty AD. Characterization and reproducibility of HepG2 hanging drop spheroids toxicology *in vitro*. Toxicol *In Vitro*. 2018;50:86-94. DOI: 10.1016/j.tiv.2018.02.013.

[43] Bai Y, Yeung E, Lui C, Ong CS, Pitaktong I, Huang C, Inoue T, Matsushita H, Ma C, Hibino NA. Net Mold-based Method of Scaffold-free Three-Dimensional Cardiac Tissue Creation. J. Vis. Exp. 2018; 138, e58252. DOI:10.3791/58252.

[44] Raghavan S, Mehta P, Horst EN, Ward MR, Rowley KR, Mehta G. Comparative analysis of tumor spheroid generation techniques for differential *in vitro* drug toxicity. Oncotarget. 2016;7(13):16948-61. DOI: 10.18632/oncotarget.7659.

[45] Lewis EEL, Wheadon H, Lewis N, Yang J, Mullin M, Hursthouse A, Stirling D, Dalby MJ, Berry CC. A Quiescent, Regeneration-Responsive Tissue Engineered Mesenchymal Stem Cell Bone Marrow Niche Model via Magnetic Levitation. ACS Nano. 2016; 10, 9, 8346-8354. DOI:10.1021/acsnano.6b02841.

[46] Lewis NS, Lewis EE, Mullin M, Wheadon H, Dalby MJ, Berry CC. Magnetically levitated mesenchymal stem cell spheroids cultured with a collagen gel maintain phenotype and quiescence. J Tissue Eng. 2017; 8:2041731417704428. DOI: 10.1177/2041731417704428.

[47] Souza GR, Molina JR, Raphael RM, Ozawa MG, Stark DJ, Levin CS, Bronk LF, Ananta JS, Mandelin J, Georgescu MM, Bankson JA, Gelovani JG, Killian TC, Arap W, Pasqualini R. Three-dimensional tissue culture based on magnetic cell levitation. Nat Nanotechnol. 2010; (4):291-296. DOI: 10.1038/nnano.2010.23.

[48] Dekkers J, Berkers G, Kruisselbrink E, Vonk A, de Jonge H, Janssens HM, Bronsveld, I., van de Graaf EA, Nieuwenhuis EE, Houwen RH, Vleggaar FP, Escher JC, de Rijke YB, Majoor CJ, Heijerman HG, de Winter-de Groot KM, Clevers H, van der Ent CK, Beekman JM. Characterizing responses to CFTR-modulating drugs using rectal organoids derived from subjects with cystic fibrosis. SciTransl Med. 2016; 8:344ra84. DOI:10.1126/scitranslmed.aad8278.

[49] Seidlitz T, Koo BK,Stange DE. Gastric organoids—an *in vitro* model system for the study of gastric development and road to personalized medicine. Cell Death Differ. 2021; 28, 68-83. DOI: 10.1038/s41418-020-00662-2.

[50] Cunniff B, Druso JE, van der Velden JL. Lung organoids: advances in generation and 3D-visualization. HistochemCell Biol. 2021; 155, 301-308. DOI: org/10.1007/s00418-020-01955-w.

[51] Prior N, Inacio P, Huch M. Liver organoids: from basic research to therapeutic applications. Gut. 2019; 68(12):2228-2237. DOI:10.1136/gutjnl-2019-319256.

[52] Wills ES, Drenth JP. Building pancreatic organoids to aid drug development. Gut. 2017; 66(3):393-394. DOI:10.1136/gutjnl-2016-312865.

[53] Cowan CS, Renner M, De Gennaro M, Gross-Scherf B, Goldblum D, Hou Y, Munz M, Rodrigues TM, Krol J, Szikra T, Cuttat R, Waldt A, Papasaikas P, Diggelmann R, Patino-Alvarez CP, Galliker P, Spirig SE, Pavlinic D, Gerber-Hollbach N, Schuierer S, Aldin Srdanovic 8, Marton Balogh 3, Riccardo Panero 3, Akos Kusnyerik, Arnold Szabo A, Michael B Stadler MB, Selim Orgül S, Simone Picelli S, Pascal W Hasler PW, Andreas Hierlemann A, Scholl HPN, Roma G, Nigsch F, Roska B. Cell Types of the Human Retina and Its Organoids at Single-Cell Resolution. Cell. 2020; 182(6),1623-1640.e34. DOI: 10.1016/j.cell.2020.08.0.

[54] Saito Y, Onishi N, Takami, H, Seishima R, Inoue H, Hirata Y, Kameyama K, Tsuchihashi K, Sugihara E, Uchino S, Ito K, Kawakubo H, Takeuchi H, Kitagawa Y, Saya H, Nagano O. (). Development of a functional thyroid model based on an organoid culture system. Biochemical and biophysical research communications. 2018; 497(2), 783-789. DOI:10.1016/j.bbrc.2018.02.154.

[55] Cruz NM, Song X, Czerniecki SM, Gulieva RE, Churchill AJ, Kim YK,

Winston K, Tran LM, Diaz MA, Fu H, Finn LS, Pei Y, Himmelfarb J, Freedman BS. (). Organoid cystogenesis reveals a critical role of microenvironment in human polycystic kidney disease. Nature materials. 2017; 16(11), 1112-1119. DOI: 10.1038/nmat4994

[56] Bareiss PM, Metzger M, Sohn K, Rupp S, Frick JS, Autenrieth IB, Lang F, Schwarz H, Skutella T, Just L. Organotypical tissue cultures from adult murine colon as an *in vitro* model of intestinal mucosa. Histochemistry and cell biology. 2008; 129(6), 795-804. DOI: 10.1007/s00418-008-0405-z.

[57] Qian X, Nguyen HN, Jacob F, Song H, Ming GL. Using brain organoids to understand Zika virus-induced microcephaly. Development. 2017; 144(6):952-957. DOI:10.1242/dev.140707

[58] Yakavets I, Francois A, Benoit A, Merlin JL, Bezdetnaya L, Guillaume Vogin G. Advanced co-culture 3D breast cancer model for investigation of fibrosis induced by external stimuli: optimization study. Sci Rep. 2020; 10,21273. DOI:10.1038/s41598-020-78087-7.

[59] Kunz-Schughart LA, Heyder P, Schroeder J, Knuechel R. A heterologous 3-D coculture model of breast tumor cells and fibroblasts to study tumor-associated fibroblast differentiation. ExpCell Res. 2001; 266(1):74-86. DOI:10.1006/excr.2001.5210.

[60] Saraiva DP, Matias AT, Braga S, Jacinto A, Cabral MG. Establishment of a 3D Co-culture With MDA-MB-231 Breast Cancer Cell Line and Patient-Derived Immune Cells for Application in the Development of Immunotherapies. Front Oncol. 2020; 10:1543. DOI:10.3389/fonc.2020.01543.

[61] Chen S, Giannakou A, Golas J, Geles KG. Multidimensional Coculture System to Model Lung Squamous Carcinoma Progression. J. Vis. Exp. 2020; (157), e60644, DOI:10.3791/60644.

[62] Amann A, Zwierzina M, Gamerith G, Bitsche M, Huber J M, Vogel GF, Blumer M, Koeck S, Pechriggl EJ, Kelm JM, Hilbe W, Zwierzina H. Development of an Innovative 3D Cell Culture System to Study Tumour - Stroma Interactions in Non-Small Cell Lung Cancer Cells. PLoS ONE. 2014; 9(3), e92511. DOI:10.1371/journal.pone.0092511.

[63] Brodaczewska KK, Bielecka ZF, Maliszewska Olejniczak K, Szczylik C, Porta C, Bartnik E, Czarnecka AM. Metastatic renal cell carcinoma cells growing in 3D on poly-D-lysine or laminin present a stem-like phenotype and drug resistance. Oncol Rep. 2019;42(5):1878-1892. DOI:10.3892/or.2019.7321.

[64] Dolznig H, Rupp C, Puri C, Haslinger C, Schweifer N, Wieser E, Kerjaschki D, Garin-Chesa P. Modeling colon adenocarcinomas *in vitro* a 3D co-culture system induces cancer-relevant pathways upon tumor cell and stromal fibroblast interaction. Am J Pathol. 2011; 179(1):487-501. DOI:10.1016/j.ajpath.2011.03.015.

[65] Dolznig H, Rupp C, Puri C, Haslinger C, Schweifer N, Wieser E, Kerjaschki D, Garin-Chesa P. Modeling colon adenocarcinomas *in vitro* a 3D co-culture system induces cancer-relevant pathways upon tumor cell and stromal fibroblast interaction. The American journal of pathology. 2011; 179(1), 487-501. DOI:10.1016/j.ajpath.2011.03.015.

[66] Takasato M, Little MH. Making a Kidney Organoid Using the Directed Differentiation of Human Pluripotent Stem Cells. Methods Mol Biol. 2017; 1597:195-206. DOI:10.1007/978-1-4939-6949-4_14.

[67] Skaper SD, Facci L. Central Nervous System Neuron-Glia co-Culture Models and Application to Neuroprotective Agents. Methods Mol Biol. 2018; 1727:63-80. DOI:10.1007/978-1-4939-7571-6_5.

[68] Strikoudis A, Cieślak A, Loffredo L, Chen YW, Patel N, Saqi A, Lederer DJ, Snoeck HW. Organoids Derived from Human Pluripotent Stem Cells. Cell Rep. 2019; 27(12):3709-3723.e5. DOI:10.1016/j.celrep.2019.05.077.

[69] Coll M, Perea L, Boon R, Leite SB, Vallverdú J, Mannaerts I, Smout A, El Taghdouini, A, Blaya D, Rodrigo-Torres D, Graupera I, Aguilar-Bravo B, Chesne C, Najimi M, Sokal E, Lozano JJ, van Grunsven LA, Verfaillie CM, Sancho-Bru P. Generation of Hepatic Stellate Cells from Human Pluripotent Stem Cells Enables *In Vitro* Modeling of Liver Fibrosis. Cell Stem Cell. 2018; 23(1):101-113.e7. DOI:10.1016/j.stem.2018.05.027.

[70] Saadeldin IM, Kim SJ, Choi YB, Lee BC. Improvement of cloned embryos development by co-culturing with parthenotes: a possible role of exosomes/microvesicles for embryos paracrine communication. Cell Reprogram. 2014; 16(3):223-234. DOI:10.1089/cell.2014.0003.

[71] Jalili-Firoozinezhad S, Gazzaniga FS, Calamari EL, Camacho DM, Fadel CW, Bein A, Swenor B, Nestor B, Cronce MJ, Tovaglieri A, Levy O, Gregory KE, Breault DT, Cabral J, Kasper DL, Novak R, Ingber DE. A complex human gut microbiome cultured in an anaerobic intestine-on-a-chip. Nat Biomed Eng. 2019; 3(7):520-531. DOI:10.1038/s41551-019-0397-0.

[72] Murphy LS, Wickersham N, McNeil JB, Shaver CM, May AK, Bastarache JA, Ware LB. Endothelial glycocalyx degradation is more severe in patients with non-pulmonary sepsis compared to pulmonary sepsis and associates with risk of ARDS and other organ dysfunction. Ann Intensive Care. 2017; 7(1):102. DOI:10.1186/s13613-017-0325-y.

[73] Krump-Konvalinkova V, Bittinger F, Unger RE, Peters K, Lehr HA, Kirkpatrick CJ. Generation of human pulmonary microvascular endothelial cell lines. Lab Invest. 2001; 81(12):1717-1727. DOI:10.1038/labinvest.3780385.

[74] Kosuge M, Takizawa H, Maehashi H, Matsuura T, Matsufuji S. A comprehensive gene expression analysis of human hepatocellular carcinoma cell lines as components of a bioartificial liver using a radial flow bioreactor. Liver Int. 2007; 27(1):101-108. DOI:10.1111/j.1478-3231.2006.01410.x.

[75] World Economic Forum. Top Ten Emerging Technologies [Internet]. 2016. Available from:https://www.weforum.org/agenda/2016/06/top-10-emerging-technologies-2016/ [Accessed: 2016-06-23].

[76] Stucki JD, Hobi N, Galimov A, Stucki AO, Schneider-Daum N, Lehr CM, Huwer H, Frick M, Funke-Chambour M, Geiser T, Guenat OT. Medium throughput breathing human primary cell alveolus-on-chip model. Scientific reports 8(1), 2018; 14359. DOI: 10.1038/s41598-018-32523-x.

[77] Zhang M, Xu C, Jiang L, Qin J. A 3D human lung-on-a-chip model for nanotoxicity testing. Toxicol Res (Camb). 2018;11;7(6):1048-1060. DOI: 10.1039/c8tx00156a.

[78] Zamprogno P, Wüthrich S, Achenbach S,Zamprogno P, Wüthrich S, Achenbach S, Thoma G, Stucki JD, Hobi N, Schneider-Daum N, Lehr CM, Huwer H, Geiser T, Schmid RA, Guenat OT. Second-generation lung-on-a-chip with an array of stretchable alveoli made with a biological membrane. Commun Biol.

2021;4(1):168. DOI:10.1038/ s42003-021-01695-0.

[79] Lukács B., Bajza Á, Kocsis D, Csorba A, Antal I, Iván K, Laki A J, Erdő F. Skin-on-a-Chip Device for *Ex Vivo* Monitoring of Transdermal Delivery of Drugs-Design, Fabrication, and Testing. Pharmaceutics. 2019; 11(9), 445. DOI:10.3390/ pharmaceutics11090445.

[80] Deng J, Cong Y, Han X, Wei W, Lu Y, Liu T, Zhao W, Lin B, Luo Y, Zhang X. A liver-on-a-chip for hepatoprotective activity assessment. Biomicrofluidics. 2020; 14(6):064107. DOI:10.1063/5.0024767.

[81] Petrosyan A, Cravedi P, Villani V, Angeletti A, Manrique J, Renieri AR, DeFilippo RE, Perin LP, Da Sacco SA. Glomerulus-on-a-chip to recapitulate the human glomerular filtration barrier. Nat Commun. 2019; 10, 3656. Doi:10.1038/s41467-019-11577-z.

[82] Nieskens TTG, Persson M, Kelly EJ, Sjögren AK. A Multicompartment Human Kidney Proximal Tubule-on-a-Chip Replicates Cell Polarization-Dependent Cisplatin Toxicity. Drug MetabDispos. 2020; 48(12):1303-1311. DOI:10.1124/dmd.120.000098.

[83] Sakamiya M, Fang Y, Mo X, Shen J, Zhang T. A heart-on-a-chip platform for online monitoring of contractile behavior via digital image processing and piezoelectric sensing technique. Med Eng Phys. 2020; 75:36-44. DOI:10.1016/j.medengphy.2019.10.001.

[84] Zhao Y, Rafatian N, Wang EY, Feric NT, Lai B, Knee-Walden EJ, Backx PH, Radisic M. Engineering microenvironment for human cardiac tissue assembly in heart-on-a-chip platform. Matrix Biol. 2020; 85-86:189-204. DOI:10.1016/j.matbio.2019.04.001.

[85] Zbinden A, Marzi J, Schlünder K, Probst C, Urbanczyk M, Black S, Brauchle EM, Layland SL, Kraushaar U, Duffy G, Schenke-Layland K, Loskill P. Non-invasive marker-independent high content analysis of a microphysiological human pancreas-on-a-chip model. Matrix Biol. 2020; 85-86:205-220. DOI:10.1016/j.matbio.2019.06.008.

[86] Zhiyun X, Yanghui G, YuanyuanH, Encheng L, Yan W, Jianing Z. Application of a microfluidic chip-based 3D co-culture to test drug sensitivity for individualized treatment of lung cancer. Biomaterials. 2013; 34:4109-17. DOI:10.1016/j.biomaterials.2013.02.045.

[87] Huh D, Hamilton GA, Ingber DE. From 3D cell culture to organs-on-chips. Trends Cell Biol. 2011;21(12):745-754. DOI:10.1016/j.tcb.2011.09.005.

[88] Swofford CA, Van Dessel N, Forbes NS. Quorum-sensing Salmonella selectively trigger protein expression within tumors. Proc. Natl. Acad. Sci. U S A. 2015; 112:3457-3462. DOI: 10.1073/ pnas.1414558112.

[89] Ramzy GM, Koessler T, Ducrey E, McKee T, Ris F, Buchs N, Rubbia-Brandt L, Dietrich PY, Nowak-Sliwinska P. (2020). Patient-Derived *In Vitro* Models for Drug Discovery in Colorectal Carcinoma. Cancers (Basel). 2020; 12(6):1423. DOI:10.3390/cancers12061423.

[90] Fong EL, Wan X, Yang J, Morgado M, Mikos AG, Harrington DA, Navone NM, Farach-Carson MCA. A 3D *in vitro* model of patient-derived prostate cancer xenograft for controlled interrogation of *in vivo* tumor-stromal interactions. Biomaterials. 2016; 77:164-172. DOI:10.1016/j. biomaterials.2015.10.059.

[91] Liu X, Krawczyk E, A Suprynowicz F, Palechor-Ceron N, Yuan H, Dakic A, Simic V, Zheng YL, Sripadhan P, Chen C, Lu J, Hou TW, Choudhury S, Kallakury B, Tang DG, Darling T, Thangapazham R,

Timofeeva O, Dritschilo A, Randell SH, ... Schlegel R. Conditional reprogramming and long-term expansion of normal and tumor cells from human biospecimens. Nat Protoc. 2017;12(2):439-451. doi:10.1038/nprot.2016.174.

[92] Beshiri ML, Tice CM, Tran C, Nguyen HM, Sowalsky AG, Agarwal S, Jansson KH, Yang Q, McGowen KM, Yin J, Alilin AN, Karzai FH, Dahut WL, Corey E, Kelly K. A PDX/Organoid Biobank of Advanced Prostate Cancers Captures Genomic and Phenotypic Heterogeneity for Disease Modeling and Therapeutic Screening. Clin Cancer Res. 2018; 24(17):4332-4345. DOI:10.1158/1078-0432.CCR-18-0409.

[93] Aparicio S, Hidalgo M, Kung AL. Examining the utility of patient-derived xenograft mouse models. Nat Rev Cancer. 2015;15(5):311-316. DOI:10.1038/nrc3944.

[94] Wartenberg M, Hoffmann E, Schwindt H, Grünheck F, Petros J, Arnold JRS, Hescheler J, Sauer H. Reactive oxygen species-linked regulation of the multidrug resistance transporter P-glycoprotein in Nox-1 overexpressing prostate tumor spheroids. FEBS Lett. 2005; 579(20):4541-4549. DOI:10.1016/j.febslet.2005.06.078.

[95] Gaspar VM, Costa EC, Queiroz JA, Pichon C, Sousa F, Correia IJ. Folate-targeted multifunctional amino acid-chitosan nanoparticles for improved cancer therapy. Pharm Res. 2015; 32(2):562-577. DOI:10.1007/s11095-014-1486-0.

[96] Turley SJ, Cremasco V, Astarita JL. Immunological hallmarks of stromal cells in the tumour microenvironment. Nat Rev Immunol. 2015;15(11):669-682. DOI:10.1038/nri3902.

[97] Liu F, Huang J, Ning B, Liu Z, Chen S, Zhao W. Drug Discovery via Human-Derived Stem Cell Organoids. Front Pharmacol. 2016;7:334. DOI:10.3389/fphar.2016.00334.

[98] Zhiyun X, Yanghui G, Yuanyuan H, Encheng L, Yan W, Jianing Z. Application of a microfluidic chip-based 3D co-culture to test drug sensitivity for individualized treatment of lung cancer. Biomaterials. 2013;34(16):4109-4117. DOI:10.1016/j.biomaterials.2013.02.045.

[99] Ghaderi D, Zhang M, Hurtado-Ziola N, Varki A. Production platforms for biotherapeutic glycoproteins. Occurrence, impact, and challenges of non-human sialylation. Biotechnol Genet Eng Rev. 2012; 28:147-175. DOI:10.5661/bger-28-147.

[100] Kim JY, Kim YG, Lee GM. CHO cells in biotechnology for production of recombinant proteins: current state and further potential. ApplMicrobiolBiotechnol. 2012; 93(3):917-930. DOI:10.1007/s00253-011-3758-5.

[101] Wurm FM. Production of recombinant protein therapeutics in cultivated mammalian cells. Nat Biotechnol. 2004;22(11):1393-1398. DOI:10.1038/nbt1026.

[102] Dumont J, Euwart D, Mei B, Estes S, Kshirsagar R. Human cell lines for biopharmaceutical manufacturing: history, status, and future perspectives. Crit Rev Biotechnol. 2016;36(6): 1110-1122. DOI:10.3109/07388551.2015.1084266.

Chapter 4

Isolation and Expansion of Mesenchymal Stem/Stromal Cells, Functional Assays and Long-Term Culture Associated Alterations of Cellular Properties

Chenghai Li

Abstract

Mesenchymal stem cell/stromal cells (MSCs) can differentiate into a variety of cell types, including osteocytes, adipocytes and chondrocytes. MSCs are present in the multiple types of adult tissue, such as bone marrow, adipose tissue, and various neonatal birth-associated tissues. Given their self-renewal and differentiation potential, immunomodulatory and paracrine properties, and lacking major histocompatibility complex (MHC) class II molecules, MSCs have attracted much attention for stem cell-based translational medicine research. Due to a very low frequency in different types of tissue, MSCs can be isolated and expanded *in vitro* to derive sufficient cell numbers prior to the clinical applications. In this chapter, the methodology to obtain primary bone marrow-derived MSCs as well as their *in vitro* culture expansion will be described. To assess the functional properties, differentiation assays, including osteogenesis, chondrogenesis and adipogenesis, 3-D culture of MSCs and co-culture of MSCs and tumor cells are also provided. Finally, the long-term culture associated alterations of MSCs, such as replicative senescence and spontaneous transformation, will be discussed for better understanding of the use of MSCs at the early stages for safe and effective cell-based therapy.

Keywords: Mesenchymal stem cell/stromal cells, primary culture, co-culture, 3-D culture, replicative senescence, spontaneous transformation

1. Introduction

Mesenchymal stem/stromal cells (MSCs), a multipotent stem/progenitor cell type, were initially described in bone marrow by Friedenstein et al. as rapid adherence to tissue culture vessels and the discrete "fibroblast" colonies approximately 50 years ago [1, 2]. Julius Cohnheim, a German-Jewish pathologist, firstly proposed that a fibroblast-like cell population for nonhematopoietic cells in bone marrow were involved in wound repair over 150 years ago [3]. In the late 1980s, Caplan firstly coined the name "mesenchymal stem cell (MSC)" [4]. Since then, MSCs have gained much attention over the last three decades. Many laboratories focusing on MSCs have developed diverse methods to isolate and expand MSCs from a variety of tissues. While the assessment of characteristics of MSCs is necessitated in different platforms/laboratories, most researchers come to acknowledge the lack of a universally accepted criteria to define MSCs. To address this question of cell equivalence, the Mesenchymal and Tissue Stem Cell Committee of the International Society for Cellular Therapy (ISCT) proposes three minimal criteria to define MSCs [5]: property of MSC plastic adherence, the expression of specific cellular surface antigen, and capacity for trilineage mesenchymal differentiation (osteogenesis, chondrogenesis and adipogenesis).

Human MSCs from different tissues have the varied phenotypic features, the morphologic inconsistency, and heterogeneous functional behavior [6–8]. Indeed, the properties of stem cell have not been well established yet. Due to the unknown *in vivo* multipotent properties of MSCs, the issue of MSC nomenclature remains actively controversial. In 2019, ISCT MSC committee issued a position statement on nomenclature of MSCs clarifying the functional definition to emphasize the functional distinction of mesenchymal stem versus stromal cells [9].

MSCs have been considered as a promising therapeutic tool in tissue engineering and regenerative medicine. MSCs are well known to be present in almost every type of adult tissues, such as bone marrow [10–12], adipose tissue [10, 13, 14], lung [11, 15], synovial tissue [16, 17], dental pulp and periodontal ligament [18]. Notably, it has become apparent that MSCs are identified in the various human embryonic tissues, such as fetal bone marrow [19], fetal liver [20], aorta gonad-mesonephros and yolk sac [21], as well as multiple neonatal birth-associated tissues, such as placenta [10, 22, 23], amniotic and chorionic membrane [23, 24], umbilical cord tissue [10, 23–25], and umbilical cord blood [26, 27]. Therefore, different platforms/laboratories may use different type of tissue sources and methodologies for isolation and expansion of MSCs. This chapter firstly outlines protocols for standardized isolation and expansion of human bone marrow-derived MSCs (BM-MSCs), a major source of human MSCs, as well as BM-MSCs' characteristics, cryopreservation and thawing. Protocols for the preparation of MSCs derived from the other tissue types are similar to that of BM-MSCs, except tissue sample processing differentially. Human BM-MSCs are estimated at a very low frequency at approximately 0.001–0.01% of total nucleated cells [28, 29], and, therefore, human BM-MSCs are likely to be kind of difficult to isolate and harvest. This chapter will then focuses on optimal functional assays and application on the basis of our previous studies, which would be useful for researchers working with MSCs in basic research and translational and clinical applications, such as osteogenesis, chondrogenesis, adipogenesis, colony forming unit-fibroblast (CFU-F) assay, 3-D cellular co-culture, MSC homing and migration. Last but not least, the long-term culture associated alterations of MSCs' properties will be also discussed in this chapter.

2. Isolation, culture expansion, phenotypic analysis, freezing and thawing of human BM-MSCs

2.1 Materials

1. Growth medium (pH 7.1–7.5): Dulbecco's modified Eagle's medium (DMEM)-low glucose (Sigma-Aldrich, St Louis, MO), 10% fetal bovine serum (FBS, heat-inactivated) (Atlanta Biologicals, Lawrenceville, GA), 0.37 g% sodium bicarbonate, and 1% penicillin–streptomycin. Adjust growth medium to pH 7.1–7.5.

2. Ficoll-Paque (Density gradient medium) (Stem Cell Technologies, Cambridge, MA).

3. 1 × Phosphate buffered saline (PBS), pH 7.2.

4. 1 × PBS + 2% + 1 mM ethylenediaminetetraacetic acid (EDTA).

5. Trypan blue, 0.4%.

6. Freezing medium: 90% FBS and 10% (v/v) dimethyl sulphoxide (DMSO).

7. Trypsin–EDTA solution 0.25% (Sigma, St. Louis, MO).

8. RosetteSep™ Human Mesenchymal Stem Cell Enrichment Cocktail (Stem Cell Technologies, Cambridge, MA).

9. Nalgene® Cryo 1°C Freezing Container.

10. Cryogenic storage vials.

11. Suitable box for storage in liquid nitrogen.

12. Tissue culture ware: T25 flasks, T75 flasks, 100-mm tissue culture dishes, 6-well plates.

13. Hemocytometer.

14. Water bath at 37°C.

15. Biological safety cabinet class II.

16. Bench centrifuge with swinging bucket rotor.

17. Inverted phase microscope.

18. Sterile conical centrifuge tubes (15 and 50 mL).

19. Sterile cell culture plastic pipettes (2, 5, 10, 25 mL).

20. Automatic pipettor.

21. 1.5 mL eppendorf tubes.

2.2 Isolation and culture expansion of human BM-MSCs

1. Make sure BM sample, 1 × PBS + 2% FBS + 1 mM EDTA, Ficoll-Paque and centrifuge are all at room temperature.

2. Spray the sample tube with 70% ethanol and conduct the isolation of BM-MSCs in the Biological Safety cabinet.

3. For 10 mL size of BM, divide BM sample into two 50 mL conical tubes at room temperature. Add 50 μl RosetteSep Human Mesenchymal Stem Cell Enrichment Cocktail per mL of bone marrow and mix well.

4. Incubate for 20 minutes at room temperature.

5. Dilute sample with about same volume of 1 × PBS + 2% FBS + 1 mM EDTA solution; mix gently.

6. Prepare two 50 mL conical tubes with 10 mL Ficoll-Paque each tube. Layer the diluted sample on top of Ficoll-Paque. Be careful to minimize mixing of Ficoll-Paque and sample. Tilt the tube to 45 degree angle and slowly add the sample drop by drop to form a layer on top.

7. Centrifuge at 1200 × g for 10 minutes at room temperature in a swinging bucket rotor and set the centrifuge to brake on.

8. Remove the enriched cells from the Ficoll-Paque. Collect the interphase containing the mononuclear cells. Be sure not to touch the red pellet at the bottom. It is advisable to remove some of the Ficoll-Paque and a bit of upper plasma layer in order to endure complete recovery of the desired cells.

9. Wash enriched cells with the 5× volume of 1 × PBS + 2% FBS + 1 mM EDTA solution. Spin 300 × g for 10 minutes.

10. Carefully discard the supernatant and resuspend cells in 1 mL of MSC growth medium per tube. Perform a viable cell count with a hemocytometer using Trypan blue.

11. Seed cells into one T-75 flask finally for 10 mL size marrow (a final cell concentration of 0.5–1.5 × 10^6 cells/cm^2). 12 mL of MSC growth medium is supplemented.

12. Put the flask in incubator at 37°C with 5% humidified CO_2 for 48 hours to allow cells to attach.

13. After 48 hours, observe with phase contrast microscopy and then remove growth medium and non-adherent cells.

14. Wash cells twice with pre-warmed medium and add 13 mL of fresh MSC growth medium. Return the flask to the incubator.

15. Change growth medium every 3 days and observe the cellular colony forming.

16. CFU-F become in the next 3–5 days. Continue to culture until the cells reach 80% confluence in the 2 weeks.

17. Remove the medium and wash with PBS 2–3 times.

18. Add 3–4 mL pre-warmed trypsin–EDTA solution to cover cells in the flask. Return the flask to the incubator for 5 minutes.

19. Check with phase contrast microscopy. When most cells become detached, gently tap the side of the flask.

20. Add 5 mL growth medium to the flask to stop Trypsin–EDTA action. Resuspend cells by pipetting and transfer the entire cell suspension into a 15 mL conical tube.

21. Centrifuge at 400 × g for 5 minutes.

22. Remove the supernatant and resuspend the cells in 2–3 mL pre-warmed PBS. Centrifuge at 400 × g for 5 minutes.

23. Repeat step 22.

24. Harvest cells. This culture is considered as passage 0.

25. Count cells and reseed cells at an optimum density of 5000 cells/cm^2 in the appropriate tissue culture ware.

26. After 24 hours, remove the growth medium and wash the cells attached to the plate once with PBS.

27. Add the appropriate volume of fresh culture medium and incubate the cells for 2–3 days.

28. When the culture reaches 80–90% confluence, remove the culture medium and wash with PBS 2–3 times.

29. For trypsinization, add an appropriate volume of pre-warmed trypsin–EDTA solution to cover the entire cellular surface. Return the flask to the incubator for 5 minutes.

30. Observe under a phase contrast microscope. When most cells become detached, gently tap the side of the flask.

31. Add the appropriate volume of growth medium to stop Trypsin–EDTA action. Mix and collect the mixture of the entire cell suspension into a 15 mL conical tube.

32. Centrifuge at 400 × g for 5 minutes at room temperature.

33. Remove the supernatant without disturbing the cell pellet and resuspend the cells with an appropriate volume of pre-warmed PBS.

34. Wash and centrifuge at 400 × g for 5 minutes again.

35. Harvest cells. This culture is passage 1. MSCs at passage 1 can be frozen in liquid nitrogen (see the next) or continue to serially passage.

2.3 Characteristics of the expanded human BM-MSCs

The following antibodies are used in flow cytometry analysis of MSC characterization, CD29-PE, CD34-PE, CD44-PE, CD73-PE, CD90-PE, CD45-FITC, CD147-FITC, HLA DR-FITC, IgG1-PE, and IgG1-FITC (BD Biosciences). IgG1 immunoglobulin is used as isotype negative controls and passage 3 MSCs are characterized using PE or FITC conjugated antibodies against the cellular surface markers.

1. Remove the growth medium and wash cells with pre-warmed PBS twice.

2. Add the pre-warmed trypsin–EDTA solution to the flask. Return the flask to the incubator for 5 minutes.

3. When most cells become detached, gently tap the side of the flask.

4. Add growth medium to the flask. Mix and transfer the entire cell suspension into a 15 mL conical tube.

5. Centrifuge at 400 × g for 5 minutes.

6. Remove the supernatant and resuspend the cells in the pre-warmed PBS.

7. Centrifuge at 400 × g for 5 minutes.

8. Carefully discard the supernatant and resuspend cells with pre-warmed PBS.

9. Centrifuge at 400 × g for 5 minutes.

10. Harvest cells.

11. Count cells and make it a single cell suspension at concentration of $1x10^6$ cells/mL in PBS.

12. Add 250 μL cellular suspension to the appropriate number of FACS tubes.

13. Add antibodies to the FACS tubes – 10 μL per antibody per tube.

14. Incubate the FACS tubes for 20–25 minutes at room temperature on a shaker to prevent aggregation.

15. Then spin down FACS tubes at 150 rpm for 5 minutes.

16. Wash twice with PBS.

17. Add 2 mL MSC growth media.

18. Centrifuge the tubes at 150 rpm for 5 minutes.

19. Pipette off the supernatant.

20. Add 250 μL MSC growth media.

21. Run samples through the FACS machine.

2.4 Freezing of BM-MSCs

Of note, perform all following steps under sterile conditions.

1. Place freezing medium on ice.

2. Harvest the cultivated MSCs, as described above (steps 1–10 in Section 2.3).

3. Resuspend MSCs in freezing medium at a concentration of $5–10 \times 10^5$ cells/mL.

4. Transfer cells into appropriate cryogenic storage vials or tubes and close the lid.

5. Place the vials in the pre-cooled Cryo 1°C Freezing Container quickly and store at −80°C directly.

6. After 24 hours, transfer the cryogenic storage vials to the suitable boxes to liquid nitrogen for long-term storage.

2.5 Thawing of BM-MSCs

Perform all following steps under sterile conditions.

1. Place the frozen vial of MSCs rapidly into a 37°C water bath.

2. Gently swirl the cryovial until the ice in the vial has melted.

3. After thawing, transfer the entire content of the cryovial into a sterile 15 mL conical tube containing 5 mL of pre-warmed PBS.

4. Gently swirl and centrifuge at 400 × g for 5 minutes.

5. Aspirate supernatant completely.

6. Wash again.

7. Count and reseed cells at a density of 5000 cells/cm^2 in the appropriate tissue culture ware.

3. Functional analysis of expanded human MSCs

3.1 *In vitro* osteogenesis of human BM-MSCs

Osteogenic medium: MSC growth medium supplemented with 50 μM ascorbic acid phosphate (AsAP) (Wako Chemicals USA, Richmond, VA), 0.1 μM dexamethasone (Sigma-Aldrich, St Louis, MO), and 10 mM β-glycerol phosphate (Sigma-Aldrich, St Louis, MO).

Several osteogenic markers, such as alkaline phosphatase (ALP), leptin receptor, and cathepsin K, are used as the indicator of early osteogenesis [28], calcium deposition as the indicator of late-stage osteogenesis [28, 29]. ALP activity and calcium deposition are exemplified to assess osteogenic differentiation of human MSCs from passage 4 (**Figure 1**).

3.1.1 Colorimetric quantitative ALP activity

1. Collect cells from passage 4.

2. Prepare cellular suspension in growth medium.

3. Plate cellular suspension in 6-well plates at 1 × 10^5 cells/well in triplicate.

4. After 24 hours, remove the growth medium and wash cells with osteogenic medium once.

5. Add 2 mL osteogenic medium in each well.

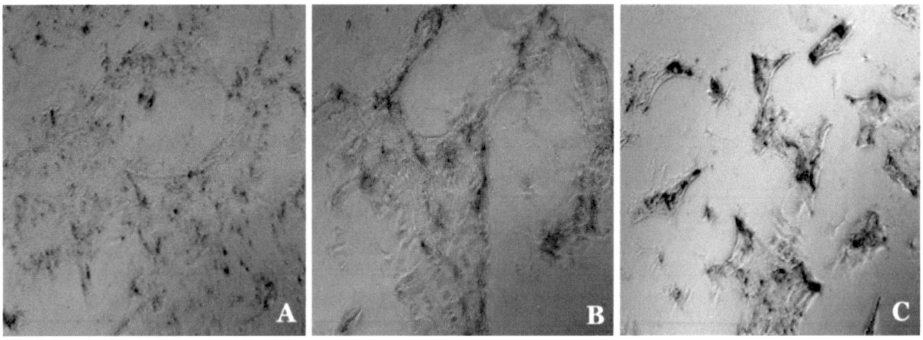

Figure 1.
Osteogenic culture of human BM-MSCs. Representative images (10×) of human BM-MSC osteogenesis on day 12 (A), day 17 (B), and day 26 during osteogenic culture.

6. Culture cells in incubator at 37°C with 5% humidified CO_2.

7. Change osteogenic medium every 3 days.

8. Measure ALP activity about 1–2 weeks after MSC osteogenetic culture [29, 30]. Aspirate medium and gently wash cells twice with PBS.

9. Add 600 μL of lysis buffer (0.5% Triton-X 100 in molecular grade ddH_2O) to each well and then scrape the cells off the surface using the end of a pipette tip.

10. Transfer lysates to centrifuge tubes. Prepare 15 mL conical tubes equal to the number of tubes maintaining lysates.

11. Dissolve the contents of a 40 mg capsule of phosphatase substrate (Sigma) in 10 mL ddH_2O. Scale up if necessary.

12. Add 500 μL of 1.5 M alkaline buffer solution (Sigma) to each tube. Add 500 μL of phosphatase substrate solution into each tube. Keep tubes in 37°C water bath.

13. Vortex samples and add 100 μL of each lysate to 15 mL tubes within 30 seconds. Fifteen minutes after the first sample is added, add 1 mL of 1 N NaOH to each tube in 30 second intervals removing tubes from water bath. The reaction will take place for 15 minutes at 37°C for each tube.

14. Prepare standard curve (Sigma) (Note: prepare 100 nmol/mL solution of p-nitrophenol by combining 100 μL of 10 mM p-nitrophenol standard solution with 9.9 mL 0.02 N NaOH).

15. Add 300 μL of standards and samples in triplicate to a 96-well plate.

16. Measure the absorbance using excitation filter of 405 nm. One enzyme unit of ALP is defined as the quantity of enzyme which produces 1 nmol p-nitrophenol per 15 minutes [29].

3.1.2 Colorimetric quantitative calcium assay

1. Aspirate or pipette out all culture medium from each well of the 6-well culture plate that contains induced or control cells to be tested.

2. Rinse the cells twice with PBS. Add 200 μL of PBS to the side of each well, not to dislodge the cells. Aspirate off the PBS and re-rinse.

3. Add 125 μL of 0.5 N HCl to each well.

4. Scrape the cells off of the surface using a cell scraper and transfer the cells and HCl to a 1.5 mL polypropylene microcentrifuge tube with a tight fitting cap.

5. Add an additional 125 μL of 0.5 N HCl to each cell to recover any cells remaining in the well, and transfer this to the appropriate tube.

6. Samples may be capped tightly and stored at −20°C if they are not to be tested immediately.

7. Extract the calcium from the cells by shaking the tubes on an orbital shaker for 4 hours at 4°C. If using frozen samples, allow extra time for samples to thaw.

8. Centrifuge the sample tubes at 500 g for 2 minutes.

9. Carefully collect the supernatant with extracted calcium, without disrupting the cell pellets, and transfer these to a new tube.

10. Prepare a standard curve with the calcium standard and determine the amount of calcium in each control and osteogenesis-induced samples. Follow the instructions provided in the Stanbio Total Calcium LiquiColor® Procedure No. 0150 (Stanbio Laboratory).

11. Three μL of sample vs. 297 μL of assay reagent (1:100 ratio for sample to reagent) for 96-well plate is used for each calcium determination. Assay reagent is mixed by equal volume of two solutions (Color Reagent and Base Reagent) provided in the kit. Distribute the assay reagent by multipipettor after adding samples. Absorbance is read at 550 nm.

12. Unused sample extract may be re-frozen for future re-assay. If the reading was out of range, the sample can be diluted with ddH_2O in a total volume of 3 μL (e.g. 1 μL of sample+2 μL of ddH_2O) and re-assayed again.

3.1.3 3-D culture of human BM-MSCs and calcium deposition measured by von Kossa staining

PuraMatrix™ hydrogel (BD Biosciences) is a 16-amino acid synthetic peptide hydrogel composed of a repeating sequence of arginine, alanine, aspartate, and alanine (RADA16) [31], which is widely used for 3-D culture.

1. 3-D culture of human BM-MSCs on PuraMatrix hydrogel as follows.

2. The stock of 1% peptide solution can be sonicated for 30 minutes to decrease its viscosity and then diluted with sterile ddH_2O to a final concentration of 0.25% (w/v).

3. 300 μL of 0.25% gel solution is loaded into each well of the 24-well plate until it is uniformly spread.

4. The gelation is initiated by slowly dripping the medium along the wall of the well. 300 μL of medium is again added carefully on top and the plate is incubated at room temperature for one hour equilibration.

5. After the peptide has assembled into hydrogel, the medium is changed two times over one hour to equilibrate the growth environment to physiological pH.

6. The equilibrated samples are stored overnight at 37°C incubator and the cells are seeded the next day.

7. Prepare the total of 4×10^4 cells suspended in MSC growth medium and 4×10^4 cells are seeded onto the hydrogel. The following day (Day 0), the medium will be replaced by osteogenic medium.

8. Von Kossa staining can be conducted at day 24 after differentiation [30]. MSCs are rinsed with the Tyrode's balanced salt solution and fixed with 10% buffered formalin (Fisher Scientific) for 30 minutes.

9. Incubated with 2% silver nitrate solution for 10 minutes in the dark.

10. Rinse with ddH_2O and expose to light for 15 minutes.

11. Bright-field images of stained samples are captured with an inverted microscope.

3.1.4 3-D cellular culture conducted by encapsulation of human MSCs in PuraMatrix™ hydrogel

1. To generate a 0.25% final concentration of PuraMatrix™ hydrogel for cells encapsulation, one part of 1% PuraMatrix™ hydrogel is diluted with same volume of sterile 20% sucrose, to reach 0.5% PuraMatrix™ hydrogel in 10% sucrose, and then mix with one part of 2× concentration of cells resuspended in 10% sucrose.

2. For 24-well plates, 300 μL of PuraMatrix mixture is loaded into each well and 300 μL of medium is layered on top of the gel. The gelation of the PuraMatrix is completed in an incubator for 60 minutes.

3. Change medium the next day.

4. Von Kossa staining can also be conducted, as described above, or collect cells as follows for other experiments.

5. Mechanically disrupt BD PureMatrix™ and cells in the well or cell culture insert by pipetting the media and gel up and down.

6. Transfer to a 15 mL conical tube.

7. Rinse out the well or cell culture insert using PBS.

8. Centrifuge at 150 × g for 5 minutes. Discard supernatant. The pellet at the bottom of the tube contains cells and BD PuraMatrix fragments.

9. Resuspend pellet in 2 mL of PBS. Spin and collect pellet again.

10. Resuspend pellet in 1 mL of trypsin–EDTA and incubate at 37°C for 5–10 minutes. This will help separate cells that are still attached to each other.

11. Add 5 mL PBS to spin cell pellet again.

12. Aspirate the supernatant (do not disturb the gel). Resuspend pellet again in 1 mL of trypsin–EDTA and incubate at 37°C for 5–10 minutes.

13. Add 5 mL PBS to spin cell pellet again.

14. Aspirate the supernatant. Carefully take out one third of the gel pellet. Do not disturb the bottom (two-third) part of the gel.

15. Wash with PBS twice.

16. Add appropriate amount of lysis buffer to perform cell lysis and collect cell sample.

3.2 3-D cell pellet culture and chondrogenesis of human BM-MSCs

Chondrogenic medium: 95% DMEM high-glucose medium (Sigma-Aldrich, St Louis, MO), 1% 1 × ITS+1 solution (BD Biosciences, San Jose, CA), 1% Pen-Strep, 100 μg/mL sodium pyruvate (Invitrogen, Carlsbad, CA), 50 μg/mL AsAP, 40 μg/mL L-proline (Sigma-Aldrich, St Louis, MO), 0.1 μM dexamethasone, and 10 ng/mL recombinant human TGF-β3 (Lonza, Walkersville, MD).

Passage 4 human BM-MSCs are used for chondrogenic differentiation. Chondrogenic differentiation is induced by chondrogenic medium (**Figure 2**).

1. In order to form a chondrogenic pellet, approximately 2.5×10^5 human BM-MSCs are centrifuged down in a 15 mL conical tube at 150 × g for 5 minutes at room temperature.

2. Five hundred μL of chondrogenic medium is used to resuspend the 2.5×10^5 cells to a final concentration of 0.5×10^6 cells/mL.

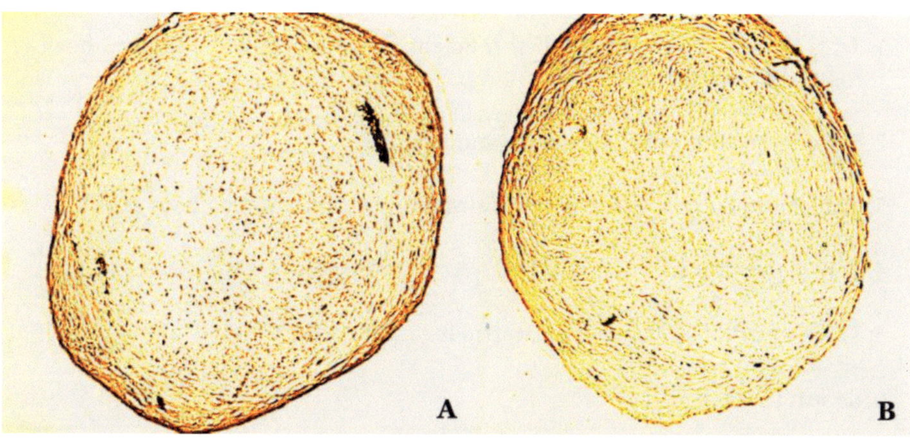

Figure 2.
Chondrogenic culture of human BM-MSCs. Representative images (10×) of immunohistochemical staining of collage II (A) and aggrecan (B) in pellet culture samples at day 17 during chondrogenesis [29].

3. Cells are centrifuged down again. Place the 15 mL conical tube in the incubator at 37°C with 5% humidified CO_2.

4. MSCs are shaped into the pellet after 24 hours incubation without disturbance.

5. The cell pellet is fed every 3 days for about 2–4 weeks and, after that, the chondrogenic pellet is harvested and sample processing is described as follows for immunohistochemistry analysis (e.g. examination of the expression levels of Collagen II, X and Aggrecan) [29, 32].

3.2.1 Sample processing for immunohistochemistry analysis

Day 1: Fixation and dehydration

1. Rinse specimen in PBS.

2. Fix samples for 45 minutes in acid-formalin at 4°C.

3. Rinse specimen in PBS twice.

4. Embed specimen in 2% agarose.

5. Transfer to vial and treat with 50% absolute ethanol for 1 hour at room temperature.

6. Transfer to 70% Absolute ethanol for 1 hour at room temperature.

7. Transfer to 95% Absolute ethanol for 1 hour at room temperature.

8. Repeat the step 7.

9. Transfer to 100% absolute ethanol for 1 hour at room temperature.

10. Leave specimen at 4°C overnight.

Day 2: Clearing and infiltration

1. Transfer to 100% CitriSolv (Fisher Scientific, catlog 22-143-975) for 1 hour at room temperature.

2. Transfer to 100% CitriSolv for 1 hour at 55°C.

3. Transfer to 1:1 mixture of CitriSolv/Micro-cut paraffin for 1 hour at 55°C.

4. Repeat the step 3.

5. Transfer to 100% Micro-cut paraffin for 1 hour at 55°C.

6. Repeat the step 5.

Day 3: Embedding

1. Transfer to 100% Micro-cut paraffin for 1–2 hours at 55°C.

2. Position specimen in Peel-Away mold with 100% paraffin.

3. Allow specimen to harden overnight.

4. Specimen may be sectioned the following day.

5. Conduct immunohistochemistry.

3.3 *In vitro* adipogenesis of human MSCs

Adipogenic medium: DMEM (1 g/L glucose), 10% FBS, 1% penicillin/streptomycin, 10 mg/mL insulin, 1 mM dexamethasone, 0.5 mM methylxanthine, and 200 mM indomethacin.

Passage 4 human BM-MSCs are used for adipogenic differentiation. Adipogenic differentiation is induced by adipogenic medium.

1. Harvest cells from passage 4, as described in the previous section.

2. Resuspend cells in adipogenic medium carefully.

3. Transfer the single cell suspension in triplicate to 6-well plates (1×10^5 cells/well).

4. Culture cells in the incubator at 37°C with 5% humidified CO_2 for 3 weeks.

5. Change medium every 2–3 days.

6. After 3 weeks, aspirate adipogenic medium and wash cells twice with PBS.

7. Cells are fixed in ice-cold methanol for 10 minutes.

8. Aspirate methanol completely and wash cells twice with ddH_2O.

9. Add Oil Red O staining reagent to the wells to stain lipid vacuoles at room temperature and mix slowly about 20 minutes on a shaker plate.

10. Aspirate Oil Red O staining reagent and wash cells twice with ddH_2O.

11. Observe and check the stained cells with phase contrast microscopy.

3.4 CFU-F assay

CFU-F assay can be used *in vitro* to evaluate the proliferation potential of MSCs. CFU-F assay is a well-established method for the quality control of MSCs' preparation. This section describes a traditional assay for CFU-F to evaluate the colony forming ability of human MSCs.

1. Collect cells from passage 4.

2. Prepare the single cell suspension in growth medium and seed cells in the 6-well plates in triplicate at three different densities, 1.5×10^5, 2.5×10^5, and 5×10^5 cells/well, in 2 mL growth medium, respectively.

3. Culture cells in the incubator at 37°C with 5% humidified CO_2 for two weeks.

4. Change medium twice each week and check with phase contrast microscopy daily to prevent overgrowth. Stop cell culture as soon as colonies are forming visibly and proceed with the Giemsa staining of CFU-F colonies on a benchtop as follows.

5. Wash the culture dishes twice with PBS.

6. Fix cells by adding 2 mL methanol to each well for 5 minutes at room temperature.

7. Gently remove the methanol and discard into the bio-hazardous waste.

8. Air dry the culture vessels and add the diluted Giemsa staining solution for 5–10 minute at room temperature.

9. Remove Giemsa staining solution and wash twice with ddH_2O.

10. Count visible colonies manually with a diameter greater than 5 mm.

3.5 Co-culture of MSCs with cancer cell line

There are various 2-D or 3-D, dyeing or not dyeing, co-culture models of human MSCs and other cell sources to study cell–cell interaction, cell proliferation, MSCs' immunomodulatory capacity, and the cellular contribution of each cell type. These methods making co-cultures of MSCs and other cells are well-established, such as co-cultures of MSCs and human peripheral blood mononuclear cells [33, 34], MSCs and T cells [35], MSCs and human hematopoietic stem cells [36], MSCs and umbilical vein endothelial cells [37]. MSC-cancer cell (PC-9) co-culture will be described in this section (**Figure 3**).

3.5.1 Preparation of MSCs with CellTracker™ Red dye (CellTracker™ CM-DiI, C7000)

1. Culture MSCs in an appropriate tissue culture ware (e.g. a 6-well plate)

2. Remove culture medium and wash with PBS twice.

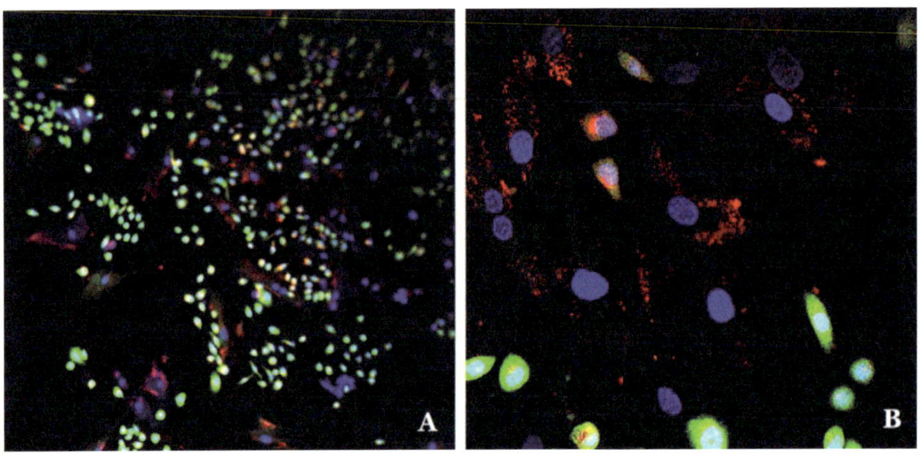

Figure 3.
Human BM-MSCs-PC-9 co-culture. Representative images ((A) 5×; (B) 20×) of co-cultures of human BM-MSCs labeled with CellTracker™ red dye and PC-9 cells transfected with copGFP, nuclei counterstained with DAPI.

3. Prepare CellTracker™ working solution in MSC growth medium. Make stock solution of lipophilic tracers in DMSO at 2 mM. Dilute the stock solution in MSC growth medium at 2 μM.

4. Add the working solution in the tissue culture vessel to cover the entire cell surface.

5. Incubate for 5 minutes or less in incubator at 37°C with 5% humidified CO_2.

6. Remove the CellTracker™ working solution.

7. Wash with MSC growth medium oncc.

8. Add MSC growth medium and return the tissue culture vessel to the incubator.

3.5.2 Preparation of PC-9 cells labeling with GFP

1. PC-9 cells labeling with copGFP (Santa Cruz, sc-108083) is performed using the lentiviral technique. Lentivirus is produced by using Lipofectamine™ 2000 (Invitrogen), according to the manufacturer's instructions. After transfection, PC-9 cells expressing copGFP may be isolated via puromycin (2 μg/mL) selection.

2. Continue to serially passage.

3. Collect PC-9 cells for cell co-culture.

3.5.3 MSC-PC-9 cell co-culture

1. Harvest MSCs with CellTrcker™ red dye and wash with PBS once.

2. Prepare the single cell suspension in growth medium at an appropriate cell concentration and transfer the cell suspension in a 6-well plate or a special chamber.

3. Culture cells in the incubator at 37°C with 5% humidified CO_2 for about 3 hours to allow cells to attach.

4. Change fresh growth medium slowly.

5. Add PC-9 cells in the 6-well plate (the same cell number of MSCs). Gently tap the side of the flask.

6. Return the 6-well plate to the incubator and culture cell overnight.

7. Check the cell co-culture under the contrast microscope and image under microscopy.

3.6 *In vitro* migration assay

The intercellular communication can be executed through a direct cell–cell interaction or through paracrine signaling mediated by a combination of active molecules. The major signaling molecules include cytokines, growth factors, chemokines, which can be generated and expressed in a wide variety of cell types including tumor cells and MSCs in response to multiple signals such as inflammatory or tumor microenvironment. Circulating MSCs are driven by such signaling molecules to home and subsequently migrate into the sites of tissue injury or disease. It is critical for the ability of MSCs to migrate and identify the injury sites for tissue repair and regeneration. Clinical data are still lacking for MSCs' homing and distribution of transplanted MSCs in the body, albeit a large number of *in vivo* studies are conducted on homing and migration pathways of MSCs for targeted stem cell-based therapies.

There are different approaches for improvement of MSC homing and migration. In this section, *in vitro* migration capacity of human BM-MSCs is evaluated by using an 8 μm-pore transwell chamber inserts (Corning).

1. Harvest MSCs and prepare cell suspension in the serum-free medium.

2. Transfer the cell suspension to the upper layer of a transwell insert at a density of 4×10^4 cells/cm^2 and allow cells to migrate to the lower compartment containing MSC growth medium overnight in the incubator at 37°C with 5% humidifies CO_2.

3. Cells from the upper chamber of transwell are migrated. Gently scrape the MSCs using the cotton swab at the upper layer of the membrane.

4. The migrated MSCs at the lower layer are stained with 0.1% crystal violet.

5. Check the cells and image under a light microscope.

6. Count the number of stained MSCs manually.

4. *In vitro* long-term culture associated alterations of MSCs

4.1 Morphological and immunophenotypic alterations of MSCs

It is well known that MSCs demonstrate biological alterations in the course of *in vitro* long-term culture. Different tissue derived MSCs may present different morphological and immunophenotypic characteristics in the expansion culture. At present, there is lack of a unifying definition for the "passage" of MSCs. Morphological changes are continuous during the long-term culture and expansion of BM-MSCs [38–40], which display a fibroblast-like appearance at early passages while the flattened and larger morphology as well as a visible increase of cellular granularity in late passages. For example, one previous study showed that human BM-MSCs were consistent with a morphological appearance from passage 1 to passage 6–8 and beyond that period such cells became large and flat [40].

During further cultivation, MSCs demonstrate the altered common immunological surface markers. Comparison of the early and late passages of BM-MSCs reveals that no differences are observed between passage 2 and 6 MSCs in expression of CD44, CD90, CD105, HLA-ABC, and HLA-DR, while CD106 is downregulated in MSCs of passage 6 [41]. Research has also reported that the expression pattern of the common surface markers maintains consistently with consecutive passaging up to passage 8 of BM-MSCs [40]. In contrast, the positive expression of the common surface markers such as CD73, CD90 and CD105 presents at the passage 30 of human adipose-derived MSCs (AD-MSCs) [42] and human umbilical cord MSCs (UC-MSCs) [43].

4.2 Alterations of proliferation and differentiation of MSCs

MSCs exhibit a high proliferation rare at lower passage and, however, the rapid growth kinetics decrease gradually with consecutive cell passaging. A linear correlation is observed between cumulative population doubling and days in culture up to passage 6–8 of human BM-MSCs and the passage-dependent decrease in the proliferation rate is also observed beyond that period [40]. A reduction in the proliferation in the course of long-term cultivation has been reported in human dental pulp tissue-derived MSCs [44] and human tonsil-derived MSCs [45].

The differentiation ability of human BM-MSCs vary in long-term culture manifested by the significant reduction in expression levels of the osteogenic markers, such as ALP and osteocalcin, and adipogenic markers, such as fatty acid binding protein-4 and lipoprotein lipase at the late passages [45]. It has been reported that 25% samples of BM-MSCs from different donors in the 8th passage and the 20% in the 10th passage lost their osteogenic differentiation potential [46]. Similarly, 10% BM-MSC samples in the 6th passage, the 50% in the 8th passage, and the 60% in the 10th passage also lost their adipogenic differentiation [46]. In contrast, an *in vitro* differentiation study has also reported that the potential of adipogenesis decreases in higher passages (from the 5th passage) whereas the propensity for osteogenesis increases in the long-term culture [39].

4.3 Replicative senescence during long-term culture expansion of MSCs

Replicative senescence is known as the irreversible growth arrest of the mitotic cells and is induced by telomere shortening. The expression of senescence markers such as senescence-associated β-galactosidase, heterochromatin protein-1, and p16INK4a increase during aging [47, 48]. Molecular damage and epigenetic alterations occur in aging stem cells [49], which can result in the impairment of stem cell function.

There are various signaling pathways involved in the senescence of MSCs, including oxidative damage [50, 51], age-related defects [52], and senescence associated up-regulation of microRNAs [53]. MSC senescence can be observed with long-term *in vitro* cultivation [54, 55], thus suggesting that a certain proportion of MSCs may undergo senescence during culture expansion. *In vitro* long-term culture of MSCs can induce continuous changes in gene expression [39, 56]. The expression levels of the senescence related genes, such as p16, p21 and p53, increase gradually in MSCs in the course of *in vitro* culture expansion [57]. DNA-methylation changes in MSCs during long-term culture have been investigated as an important epigenomic feature of replicative senescence of MSCs [58–60]. DNA-methylation changes may affect the proliferation and differentiation of MSCs. Differential methylation patterns of gene and miRNAs show between early-passage (passage 5) and late-passage (passage 15) MSCs [60]. Some genes that are hypermethylated at passage 5 present the lower mRNA expression than does these hypermethylated at passage 15 and vise versa [60].

Senescent cells secrete a complex combination of interleukins, chemokines, growth factors, proinflammatory/inflammatory cytokines, which compose the senescence-associated secretory phenotype (SASP) [61, 62]. One previous report has shown that conditioned medium (CM) collected from senescent BM-MSC culture at passage 10 is able to trigger senescence in young cells [63]. The key factors of senescent MSC CM needed for triggering senescence in the young MSCs have been characterized as insulin-like growth factor binding proteins 4 and 7, which are linked to cellular senescence and apoptosis [63]. Similarly, monocyte chemoattractant proten-1 (MCP-1), as a dominant component of the SASP, is markedly increased in the conditioned medium of the late-phase MSCs and MCP-1 treatment significantly increase the senescence phenotypes of umbilical cord blood-derived MSCs via its cognate receptor chemokine receptor 2 signaling cascade [64]. Senescence-associated changes are observed in the metabolome of MSCs during replicative senescence, including down-regulation of nicotinamide ribonucleotide and up-regulation of orotic acid, which may be used to monitor the cellular senescent state during culture expansion of MSCs [65].

4.4 *In vitro* long-term culture associated spontaneous transformation of MSCs

Sarcoma represents a very heterogeneous group of relatively rare tumors and a variety of different studies have investigated to support the MSC origin of sarcoma. There are a number of cellular and molecular mechanisms of MSC transformation for better understanding of MSCs' contribution to sarcomagenesis [66]. The majority of published research articles indicate that various sarcoma types have been shown MSCs' origin. Several group have reported spontaneous transformation in human and murine MSCs after long term culture [67–70]. For example, one study has reported that murine BM-MSCs are spontaneously transformed at passage 29 under standard conditions and that these transformed MSCs are able to generate fibrosarcoma in immunocompromised mice [70]. Accumulated chromosomal abnormalities, such as chromosome instability, chromosomal imbalances and aneuploidy, are suggested to be associated with the transformation of BM MSCs [70]. Indeed, chromosomal aberrations (chromosomal level) in *in vitro* cultures of human MSCs have been reported in previous studies, including human BM-MSCs after passage 4 [71], human AD-MSCs from passage 5 [72], and UC-MSCs from passage 5 [73].

There are also studies that have not detected the transformation of MSCs in long-term culture [74–76]. One previous study reports that human BM-MSCs do not undergo malignant transformation after long-term *in vitro* culture for up to 44 weeks and these cells maintain a normal karyotype [75]. In agree with the previous report [75], another study has described the occurrence of aneuploidy in cultivated human MSCs without evidence of transformation either *in vitro* or *in vivo* [76]. In addition, Røsland G.V. *et al* have reported that human BM-MSC spontaneous transformation phenomenon occurred in consequence of the cross-contamination between the transformed human MSCs and human cancer cells [77]. To date, there is no solid evidence for the transformation of different sarcoma subtypes from MSCs and it leaves an uncertainty for MSCs with the ability to spontaneously transform.

5. Conclusion and perspective

MSCs provide huge opportunities in translational medicine for treatment of a range of diseases or medical conditions. MSCs are multipotent stem cells and such cells can be isolated from various tissues including bone marrow, a major source of human MSCs. Given that a large number of MSCs are required for the clinical application, *in vitro* expansion of MSCs is critical. However, MSCs at higher passage could lead to the culture-associated alterations, such as cellular morphology, immunological surface markers, proliferation, differentiation, and cell genetics. Due to *in vitro* long-term culture associated spontaneous transformation of MSCs, the safety of MSC therapy remains the major concerns. Human MSCs from various tissues present the varies biological properties. At present, consensus is lacking regarding materials and culture protocols, culture conditions, supplement of growth factors, freezing and thawing (e.g. media), and functional assays. Therefore, standardizing the procedures for preparation of MSCs would be needed for pre-clinical studies and clinical application of MSCs.

Acknowledgements

This work was supported in part by Henan Provincial Engineering Research Center for Immune Cell and Stem Cell Treatment, China and Henan Key Laboratory of Stem Cell Differentiation and Modification, China.

Conflict of interest

No competing interests for this work.

Abbreviations

AD-MSCs	Adipose-derived mesenchymal stem cell/stromal cells
ALP	Alkaline phosphatase
AsAP	Ascorbic acid phosphate
BM-MSCs	Bone marrow-derived MSCs
CM	Conditioned medium
CFU-F	Colony forming unit-fibroblast
DMEM	Dulbecco's modified Eagle's medium
DMSO	Dimethyl sulphoxide
EDTA	Ethylenediaminetetraacetic acid
FBS	Fetal bovine serum
ISCT	International Society for Cellular Therapy
MCP-1	Monocyte chemoattractant proten-1
MHC	Major histocompatibility complex
MSCs	Mesenchymal stem cell/stromal cells
PBS	Phosphate buffered saline
SASP	Senescence-associated secretory phenotype
UC-MSCs	Umbilical cord mesenchymal stem cell/stromal cells

Author details

Chenghai Li
Stem Cell Program of Clinical Research Center, People's Hospital of Zhengzhou University, Zhengzhou, China

*Address all correspondence to: chenghai_li@yahoo.com

References

[1] Friedenstein AJ, Piatetzky-Shapiro II, Petrakova KV. Osteogenesis in transplants of bone marrow cells. J Embryol Exp Morphol. 1966;16:381-390.

[2] Friedenstein AJ, Chailakhjan RK, Lalykina KS. The development of fibroblast colonies in monolayer cultures of guinea-pig bone marrow and spleen cells. Cell Tissue Kinet. 1970;3:393-403. DOI: 10.1111/j.1365-2184.1970.tb00347.x

[3] Prockop DJ. Marrow stromal cells as stem cells for nonhematopoietic tissues. Science. 1997;276(5309):71-74. DOI: 10.1126/science.276.5309.71

[4] Caplan AI. Mesenchymal stem cells. J Orthop Res. 1991;9:641-650. doi: 10.1002/jor.1100090504

[5] Dominici M, Le Blanc K, Mueller I, Slaper-Cortenbach I, Marini F, Krause D, Deans R, Keating A, Prockop Dj, Horwitz E. Minimal criteria for defining multipotent mesenchymal stromal cells. The International Society for Cellular Therapy position statement. Cytotherapy. 2006;8:315-317. doi: 10.1080/14653240600855905

[6] Kozlowska U, Krawczenko A, Futoma K, Jurek T, Rorat M, Patrzalek D, Klimczak A. Similarities and differences between mesenchymal stem/progenitor cells derived from various human tissues. World J Stem Cells. 2019;11:347-374. doi: 10.4252/wjsc.v11.i6.347

[7] Melief SM, Zwaginga JJ, Fibbe WE, Roelofs H. Adipose tissue-derived multipotent stromal cells have a higher immunomodulatory capacity than their bone marrow-derived counterparts. Stem Cells Transl Med. 2013;2:455-463. doi: 10.5966/sctm.2012-0184

[8] Phinney DG, Sensebé L. Mesenchymal stromal cells: misconceptions and evolving concepts. Cytotherapy. 2013;15:140-145. doi: 10.1016/j.jcyt.2012.11.005

[9] Viswanathan S, Shi Y, Galipeau J, Krampera M, Leblanc K, Martin I, Nolta J, Phinney DG, Sensebe L. Mesenchymal stem versus stromal cells: International Society for Cell & Gene Therapy (ISCT®) Mesenchymal Stromal Cell committee position statement on nomenclature. Cytotherapy. 2019;21:1019-1024. doi: 10.1016/j.jcyt.2019.08.002

[10] Du WJ, Chi Y, Yang ZX, Li ZJ, Cui JJ, Song BQ, Li X, Yang SG, Han ZB, Han ZC. Heterogeneity of proangiogenic features in mesenchymal stem cells derived from bone marrow, adipose tissue, umbilical cord, and placenta. Stem Cell Res Ther. 2016;7:163. doi: 10.1186/s13287-016-0418-9

[11] Ricciardi M, Malpeli G, Bifari F, Bassi G, Pacelli L, Nwabo Kamdje AH, Chilosi M, Krampera M. Comparison of epithelial differentiation and immune regulatory properties of mesenchymal stromal cells derived from human lung and bone marrow. PLoS One. 2012;7:e35639. doi: 10.1371/journal.pone.0035639

[12] Drela K, Stanaszek L, Snioch K, Kuczynska Z, Wrobel M, Sarzynska S, Legosz P, Maldyk P, Lukomska B. Bone marrow-derived from the human femoral shaft as a new source of mesenchymal stem/stromal cells: an alternative cell material for banking and clinical transplantation. Stem Cell Res Ther. 2020;11:262. doi: 10.1186/s13287-020-01697-5

[13] Priya N, Sarcar S, Majumdar AS, SundarRaj S. Explant culture: a simple, reproducible, efficient and economic technique for isolation of mesenchymal stromal cells from human adipose tissue and lipoaspirate. J Tissue Eng Regen

Med. 2014;8:706-716. doi: 10.1002/term.1569

[14] Nepali S, Park M, Lew H, Kim O. Comparative Analysis of Human Adipose-Derived Mesenchymal Stem Cells from Orbital and Abdominal Fat. Stem Cells Int. 2018;2018:3932615. doi: 10.1155/2018/3932615

[15] Steens J, Klar L, Hansel C, Slama A, Hager T, Jendrossek V, Aigner C, Klein D. The vascular nature of lung-resident mesenchymal stem cells. Stem Cells Transl Med. 2021;10:128-143. doi: 10.1002/sctm.20-0191

[16] Ferro T, Santhagunam A, Madeira C, Salgueiro JB, da Silva CL, Cabral JMS. Successful isolation and ex vivo expansion of human mesenchymal stem/stromal cells obtained from different synovial tissue-derived (biopsy) samples. J Cell Physiol. 2019;234:3973-3984. doi: 10.1002/jcp.27202

[17] De Bari C, Dell'Accio F, Tylzanowski P, Luyten FP. Multipotent mesenchymal stem cells from adult human synovial membrane. Arthritis Rheum. 2001;44:1928-1942. doi: 10.1002/1529-0131(200108)44:8<1928::AID-ART331>3.0.CO;2-P

[18] Lei M, Li K, Li B, Gao LN, Chen FM, Jin Y. Mesenchymal stem cell characteristics of dental pulp and periodontal ligament stem cells after *in vivo* transplantation. Biomaterials. 2014;35:6332-6343. doi: 10.1016/j.biomaterials.2014.04.071

[19] Abdulrazzak H, Moschidou D, Jones G, Guillot PV. Biological characteristics of stem cells from foetal, cord blood and extraembryonic tissues. J R Soc Interface. 2010;7 Suppl 6(Suppl 6):S689-706. doi: 10.1098/rsif.2010.0347.focus

[20] Cronwright G, Le Blanc K, Götherström C, Darcy P, Ehnman M, Brodin B. Cancer/testis antigen expression in human mesenchymal stem cells: down-regulation of SSX impairs cell migration and matrix metalloproteinase 2 expression. Cancer Res. 2005;65:2207-2215. doi: 10.1158/0008-5472.CAN-04-1882

[21] Wang XY, Lan Y, He WY, Zhang L, Yao HY, Hou CM, Tong Y, Liu YL, Yang G, Liu XD, Yang X, Liu B, Mao N. Identification of mesenchymal stem cells in aorta-gonad-mesonephros and yolk sac of human embryos. Blood. 2008;111:2436-2443. doi: 10.1182/blood-2007-07-099333

[22] Phunikom N, Boonmuen N, Kheolamai P, Suksen K, Manochantr S, Tantrawatpan C, Tantikanlayaporn D. Andrographolide promotes proliferative and osteogenic potentials of human placenta-derived mesenchymal stem cells through the activation of Wnt/β-catenin signaling. Stem Cell Res Ther. 2021;12:241. doi: 10.1186/s13287-021-02312-x

[23] Araújo AB, Salton GD, Furlan JM, Schneider N, Angeli MH, Laureano ÁM, Silla L, Passos EP, Paz AH. Comparison of human mesenchymal stromal cells from four neonatal tissues: Amniotic membrane, chorionic membrane, placental decidua and umbilical cord. Cytotherapy. 2017;19:577-585. doi: 10.1016/j.jcyt.2017.03.001

[24] Ma J, Wu J, Han L, Jiang X, Yan L, Hao J, Wang H. Comparative analysis of mesenchymal stem cells derived from amniotic membrane, umbilical cord, and chorionic plate under serum-free condition. Stem Cell Res Ther. 2019;10:19. doi: 10.1186/s13287-018-1104-x

[25] Semenova E, Grudniak MP, Machaj EK, Bocian K, Chroscinska-Krawczyk M, Trochonowicz M, Stepaniec IM, Murzyn M, Zagorska KE, Boruczkowski D, Kolanowski TJ,

Oldak T, Rozwadowska N. Mesenchymal Stromal Cells from Different Parts of Umbilical Cord: Approach to Comparison & Characteristics. Stem Cell Rev Rep. doi: 10.1007/s12015-021-10157-3

[26] Zhao D, Liu L, Chen Q, Wang F, Li Q, Zeng Q, Huang J, Luo M, Li W, Zheng Y, Liu T. Hypoxia with Wharton's jelly mesenchymal stem cell coculture maintains stemness of umbilical cord blood-derived CD34$^+$ cells. Stem Cell Res Ther. 2018;9:158. doi: 10.1186/s13287-018-0902-5

[27] Amati E, Perbellini O, Rotta G, Bernardi M, Chieregato K, Sella S, Rodeghiero F, Ruggeri M, Astori G. High-throughput immunophenotypic characterization of bone marrow- and cord blood-derived mesenchymal stromal cells reveals common and differentially expressed markers: identification of angiotensin-converting enzyme (CD143) as a marker differentially expressed between adult and perinatal tissue sources. Stem Cell Res Ther. 2018;16;9:10. doi: 10.1186/s13287-017-0755-3

[28] Ilmer M, Karow M, Geissler C, Jochum M, Neth P. Human osteoblast-derived factors induce early osteogenic markers in human mesenchymal stem cells. Tissue Eng Part A. 2009;15:2397-2409. doi: 10.1089/ten.tea.2008.0427

[29] Li C, Sunderic K, Nicoll SB, Wang S. Downregulation of Heat Shock Protein 70 Impairs Osteogenic and Chondrogenic Differentiation in Human Mesenchymal Stem Cells. Sci Rep. 2018;8:553. doi: 10.1038/s41598-017-18541-1

[30] Chen J, Shi ZD, Ji X, Morales J, Zhang J, Kaur N, Wang S. Enhanced osteogenesis of human mesenchymal stem cells by periodic heat shock in self-assembling peptide hydrogel. Tissue Eng Part A. 2013;19:716-728. doi: 10.1089/ten.TEA.2012.0070

[31] Moradi F, Bahktiari M, Joghataei MT, Nobakht M, Soleimani M, Hasanzadeh G, Fallah A, Zarbakhsh S, Hejazian LB, Shirmohammadi M, Maleki F. BD PuraMatrix peptide hydrogel as a culture system for human fetal Schwann cells in spinal cord regeneration. J Neurosci Res. 2012;90:2335-2348. doi: 10.1002/jnr.23120

[32] Chen J, Li C, Wang S. Periodic heat shock accelerated the chondrogenic differentiation of human mesenchymal stem cells in pellet culture. PLoS One. 2014;9:e91561. doi: 10.1371/journal.pone.0091561

[33] Luetzkendorf J, Nerger K, Hering J, Moegel A, Hoffmann K, Hoefers C, Mueller-Tidow C, Mueller LP. Cryopreservation does not alter main characteristics of Good Manufacturing Process-grade human multipotent mesenchymal stromal cells including immunomodulating potential and lack of malignant transformation. Cytotherapy. 2015;17:186-198. doi: 10.1016/j.jcyt.2014.10.018

[34] Waterman RS, Tomchuck SL, Henkle SL, Betancourt AM. A new mesenchymal stem cell (MSC) paradigm: polarization into a pro-inflammatory MSC1 or an Immunosuppressive MSC2 phenotype. PLoS One. 2010;5:e10088. doi: 10.1371/journal.pone.0010088

[35] Marinescu CI, Preda MB, Burlacu A. A procedure for *in vitro* evaluation of the immunosuppressive effect of mouse mesenchymal stem cells on activated T cell proliferation. Stem Cell Res Ther. 2021;12:319. doi: 10.1186/s13287-021-02344-3

[36] Jing D, Fonseca AV, Alakel N, Fierro FA, Muller K, Bornhauser M, Ehninger G, Corbeil D, Ordemann R. Hematopoietic stem cells in co-culture with mesenchymal stromal cells--modeling the niche compartments

in vitro. Haematologica. 2010;95: 542-550. doi: 10.3324/haematol. 2009.010736

[37] Joshi A, Xu Z, Ikegami Y, Yamane S, Tsurashima M, Ijima H. Co-culture of mesenchymal stem cells and human umbilical vein endothelial cells on heparinized polycaprolactone/gelatin co-spun nanofibers for improved endothelium remodeling. Int J Biol Macromol. 2020;151:186-192. doi: 10.1016/j.ijbiomac.2020.02.163

[38] Redaelli S, Bentivegna A, Foudah D, Miloso M, Redondo J, Riva G, Baronchelli S, Dalprà L, Tredici G. From cytogenomic to epigenomic profiles: monitoring the biologic behavior of *in vitro* cultured human bone marrow mesenchymal stem cells. Stem Cell Res Ther. 2012;3:47. doi: 10.1186/scrt138

[39] Wagner W, Horn P, Castoldi M, Diehlmann A, Bork S, Saffrich R, Benes V, Blake J, Pfister S, Eckstein V, Ho AD. Replicative senescence of mesenchymal stem cells: a continuous and organized process. PLoS One. 2008;3:e2213. doi: 10.1371/journal. pone.0002213

[40] Khoo ML, Shen B, Tao H, Ma DD. Long-term serial passage and neuronal differentiation capability of human bone marrow mesenchymal stem cells. Stem Cells Dev. 2008;17:883-896. doi: 10.1089/scd.2007.0185

[41] Halfon S, Abramov N, Grinblat B, Ginis I. Markers distinguishing mesenchymal stem cells from fibroblasts are downregulated with passaging. Stem Cells Dev. 2011;20:53-66. doi: 10.1089/ scd.2010.0040

[42] Danisovic L, Oravcova L, Krajciova L, Varchulova Novakova Z, Bohac M, Varga I, Vojtassak J. Effect of long-term culture on the biological and morphological characteristics of human adipose tissue-derived stem Cells. J Physiol Pharmacol. 2017;68:149-158.

[43] Wang Y, Zhang Z, Chi Y, Zhang Q, Xu F, Yang Z, Meng L, Yang S, Yan S, Mao A, Zhang J, Yang Y, Wang S, Cui J, Liang L, Ji Y, Han ZB, Fang X, Han ZC. Long-term cultured mesenchymal stem cells frequently develop genomic mutations but do not undergo malignant transformation. Cell Death Dis. 2013;4:e950. doi: 10.1038/cddis.2013.480

[44] Wang H, Zhong Q, Yang T, Qi Y, Fu M, Yang X, Qiao L, Ling Q, Liu S, Zhao Y. Comparative characterization of SHED and DPSCs during extended cultivation *in vitro*. Mol Med Rep. 2018;17:6551-6559. doi: 10.3892/ mmr.2018.8725

[45] Yu Y, Park YS, Kim HS, Kim HY, Jin YM, Jung SC, Ryu KH, Jo I. Characterization of long-term *in vitro* culture-related alterations of human tonsil-derived mesenchymal stem cells: role for CCN1 in replicative senescence-associated increase in osteogenic differentiation. J Anat. 2014;225:510-518. doi: 10.1111/joa.12229

[46] Gharibi B, Hughes FJ. Effects of medium supplements on proliferation, differentiation potential, and *in vitro* expansion of mesenchymal stem cells. Stem Cells Transl Med. 2012;1:771-782. doi: 10.5966/sctm.2010-0031

[47] Sharpless NE, DePinho RA. How stem cells age and why this makes us grow old. Nat Rev Mol Cell Biol. 2007;8:703-713. doi: 10.1038/nrm2241

[48] Liu L, Rando TA. Manifestations and mechanisms of stem cell aging. J Cell Biol. 2011;193: 257-266. doi: 10.1083/jcb.201010131

[49] Ermolaeva M, Neri F, Ori A, Rudolph KL. Cellular and epigenetic drivers of stem cell ageing. Nat Rev Mol Cell Biol. 2018;19:594-610. doi: 10.1038/ s41580-018-0020-3

[50] Alves H, Munoz-Najar U, De Wit J, Renard AJ, Hoeijmakers JH, Sedivy JM,

Van Blitterswijk C, De Boer J. A link between the accumulation of DNA damage and loss of multi-potency of human mesenchymal stromal cells. J Cell Mol Med. 2010;14:2729-2738. doi: 10.1111/j.1582-4934.2009.00931.x

[51] Ko E, Lee KY, Hwang DS. Human umbilical cord blood-derived mesenchymal stem cells undergo cellular senescence in response to oxidative stress. Stem Cells Dev. 2012;21:1877-1886. doi: 10.1089/scd.2011.0284

[52] Stolzing A, Jones E, McGonagle D, Scutt A. Age-related changes in human bone marrow-derived mesenchymal stem cells: consequences for cell therapies. Mech Ageing Dev. 2008;129:163-173. doi: 10.1016/j.mad.2007.12.002

[53] Cakouros D, Gronthos S. The changing epigenetic landscape of Mesenchymal Stem/Stromal Cells during aging. Bone. 2020;137:115440. doi: 10.1016/j.bone.2020.115440

[54] Bonab MM, Alimoghaddam K, Talebian F, Ghaffari SH, Ghavamzadeh A, Nikbin B. Aging of mesenchymal stem cell *in vitro*. BMC Cell Biol. 2006;7:14. doi: 10.1186/1471-2121-7-14

[55] Turinetto V, Vitale E, Giachino C. Senescence in human mesenchymal stem cells: Functional changes and implications in stem cell-based therapy. Int J Mol Sci. 2016;17:1164. doi: 10.3390/ijms17071164

[56] Wagner W, Bork S, Lepperdinger G, Joussen S, Ma N, Strunk D, Koch C. How to track cellular aging of mesenchymal stromal cells? Aging (Albany NY). 2010;2:224-230. doi: 10.18632/aging.100136

[57] Cheng H, Qiu L, Ma J, Zhang H, Cheng M, Li W, Zhao X, Liu K. Replicative senescence of human bone marrow and umbilical cord derived mesenchymal stem cells and their differentiation to adipocytes and osteoblasts. Mol Biol Rep. 2011;38:5161-5168. doi: 10.1007/s11033-010-0665-2

[58] Bork S, Pfister S, Witt H, Horn P, Korn B, Ho AD, Wagner W. DNA methylation pattern changes upon long-term culture and aging of human mesenchymal stromal cells. Aging Cell. 2010;9:54-63. doi: 10.1111/j.1474-9726.2009.00535.x

[59] Schellenberg A, Lin Q, Schüler H, Koch CM, Joussen S, Denecke B, Walenda G, Pallua N, Suschek CV, Zenke M, Wagner W. Replicative senescence of mesenchymal stem cells causes DNA-methylation changes which correlate with repressive histone marks. Aging (Albany NY). 2011;3:873-888. doi: 10.18632/aging.100391

[60] Choi MR, In YH, Park J, Park T, Jung KH, Chai JC, Chung MK, Lee YS, Chai YG. Genome-scale DNA methylation pattern profiling of human bone marrow mesenchymal stem cells in long-term culture. Exp Mol Med. 2012;44:503-512. doi: 10.3858/emm.2012.44.8.057

[61] Herranz N, Gil J. Mechanisms and functions of cellular senescence. J Clin Invest. 2018;128:1238-1246. doi: 10.1172/JCI95148

[62] Di Micco R, Krizhanovsky V, Baker D, d'Adda di Fagagna F. Cellular senescence in ageing: from mechanisms to therapeutic opportunities. Nat Rev Mol Cell Biol. 2021;22:75-95. doi: 10.1038/s41580-020-00314-w

[63] Severino V, Alessio N, Farina A, Sandomenico A, Cipollaro M, Peluso G, Galderisi U, Chambery A. Insulin-like growth factor binding proteins 4 and 7 released by senescent cells promote premature senescence in mesenchymal stem cells. Cell Death Dis. 2013;4:e911. doi: 10.1038/cddis.2013.445

[64] Jin HJ, Lee HJ, Heo J, Lim J, Kim M, Kim MK, Nam HY, Hong GH, Cho YS, Choi SJ, Kim IG, Shin DM, Kim SW. Senescence-associated MCP-1 secretion is dependent on a decline in BMI1 in human mesenchymal stromal cells. Antioxid Redox Signal. 2016;24:471-485. doi: 10.1089/ars.2015.6359

[65] Fernandez-Rebollo E, Franzen J, Goetzke R, Hollmann J, Ostrowska A, Oliverio M, Sieben T, Rath B, Kornfeld JW, Wagner W. Senescence-associated metabolomic phenotype in primary and iPSC-derived mesenchymal stromal cells. Stem Cell Reports. 2020; 14: 201-209. doi: 10.1016/j.stemcr.2019.12.012

[66] Li C, Zhao H, Wang B. Mesenchymal stem/stromal cells: Developmental origin, tumorigenesis and translational cancer therapeutics. Transl Oncol. 2021;14:100948. doi: 10.1016/j.tranon.2020.100948

[67] Tang Q, Chen Q, Lai X, Liu S, Chen Y, Zheng Z, Xie Q, Maldonado M, Cai Z, Qin S, Ho G, Ma L. Malignant transformation potentials of human umbilical cord mesenchymal stem cells both spontaneously and via 3-methycholanthrene induction. PLoS One. 2013;8:e81844. doi: 10.1371/journal.pone.0081844

[68] Pan Q, Fouraschen SM, de Ruiter PE, Dinjens WN, Kwekkeboom J, Tilanus HW, van der Laan LJ. Detection of spontaneous tumorigenic transformation during culture expansion of human mesenchymal stromal cells. Exp Biol Med (Maywood). 2014;239:105-115. doi: 10.1177/1535370213506802

[69] Li H, Fan X, Kovi RC, Jo Y, Moquin B, Konz R, Stoicov C, Kurt-Jones E, Grossman SR, Lyle S, Rogers AB, Montrose M, Houghton J. Spontaneous expression of embryonic factors and p53 point mutations in aged mesenchymal stem cells: a model of age-related tumorigenesis in mice. Cancer Res. 2007;67:10889-10898. doi: 10.1158/0008-5472.CAN-07-2665

[70] Miura M, Miura Y, Padilla-Nash HM, Molinolo AA, Fu B, Patel V, Seo BM, Sonoyama W, Zheng JJ, Baker CC, Chen W, Ried T, Shi S. Accumulated chromosomal instability in murine bone marrow mesenchymal stem cells leads to malignant transformation. Stem Cells. 2006;24:1095-1103. doi: 10.1634/stemcells.2005-0403

[71] Binato R, de Souza Fernandez T, Lazzarotto-Silva C, Du Rocher B, Mencalha A, Pizzatti L, Bouzas LF, Abdelhay E. Stability of human mesenchymal stem cells during *in vitro* culture: considerations for cell therapy. Cell Prolif. 2013;46:10-22. doi: 10.1111/cpr.12002

[72] Froelich K, Mickler J, Steusloff G, Technau A, Ramos Tirado M, Scherzed A, Hackenberg S, Radeloff A, Hagen R, Kleinsasser N. Chromosomal aberrations and deoxyribonucleic acid single-strand breaks in adipose-derived stem cells during long-term expansion *in vitro*. Cytotherapy. 2013;15:767-781. doi: 10.1016/j.jcyt.2012.12.009

[73] Borghesi A, Avanzini MA, Novara F, Mantelli M, Lenta E, Achille V, Cerbo RM, Tzialla C, Longo S, De Silvestri A, Zimmermann LJ, Manzoni P, Zecca M, Spinillo A, Maccario R, Zuffardi O, Stronati M. Genomic alterations in human umbilical cord-derived mesenchymal stromal cells call for stringent quality control before any possible therapeutic approach. Cytotherapy. 2013;15:1362-1373. doi: 10.1016/j.jcyt.2013.06.006

[74] Chen G, Yue A, Ruan Z, Yin Y, Wang R, Ren Y, Zhu L. Human umbilical cord-derived mesenchymal stem cells do not undergo malignant transformation during long-term culturing in serum-free medium.

PLoS One. 2014;9:e98565. doi: 10.1371/journal.pone.0098565

[75] Bernardo ME, Zaffaroni N, Novara F, Cometa AM, Avanzini MA, Moretta A, Montagna D, Maccario R, Villa R, Daidone MG, Zuffardi O, Locatelli F. Human bone marrow derived mesenchymal stem cells do not undergo transformation after long-term *in vitro* culture and do not exhibit telomere maintenance mechanisms. Cancer Res. 2007;67:9142-9149. doi: 10.1158/0008-5472.CAN-06-4690

[76] Tarte K, Gaillard J, Lataillade JJ, Fouillard L, Becker M, Mossafa H, Tchirkov A, Rouard H, Henry C, Splingard M, Dulong J, Monnier D, Gourmelon P, Gorin NC, Sensebé L; Société Française de Greffe de Moelle et Thérapie Cellulaire. Clinical-grade production of human mesenchymal stromal cells: occurrence of aneuploidy without transformation. Blood. 2010;115:1549-1553. doi: 10.1182/blood-2009-05-219907

[77] Torsvik A, Røsland GV, Svendsen A, Molven A, Immervoll H, McCormack E, Lønning PE, Primon M, Sobala E, Tonn JC, Goldbrunner R, Schichor C, Mysliwietz J, Lah TT, Motaln H, Knappskog S, Bjerkvig R. Spontaneous malignant transformation of human mesenchymal stem cells reflects cross-contamination: putting the research field on track - letter. Cancer Res. 2010;70:6393-6396. doi: 10.1158/0008-5472.CAN-10-1305

Chapter 5

The Use of Stable Isotope Labeling with Amino Acids in Cell Culture (SILAC) to Study Ivermectin-Mediated Molecular Pathway Changes in Human Ovarian Cancer Cells

Na Li and Xianquan Zhan

Abstract

Stable isotope labeling with amino acids in cell culture (SILAC) was to use isotopic essential amino acids to replace the original amino acids for cell culture and passage for 8–10 generations, followed by mass spectrometry to identify proteins and the isotopic abundance difference to quantify proteins. SILAC can be used to characterize proteomic changes, and analyze protein turnover, protein interactions, and dynamic changes with quantitative accuracy, and high reproducibility. For this study, SILAC "light" (L-Lysine-2HCl [^{12}C6, ^{14}N2], L-Arginine-HCl [^{12}C6, ^{14}N4])- or "heavy" (L-Lysine-2HCl [^{13}C6, ^{15}N2], L-Arginine-HCl [^{13}C6, ^{15}N4])-labeling RPMI 1640 medium was used to culture human ovarian cancer TOV-21G cells for 10 passages, followed by the treatment of 0.1% dimethylsulfoxide for 24 h and 20 μM ivermectin for 24 h, respectively. The light- and heavy-isotope-labeled proteins were equally mixed (1:1) for digestion with trypsin. The tryptic peptide mixture was fractionated with liquid chromatography and analyzed with tandem mass spectrometry. In total, 4,447 proteins were identified in ivermectin-treated TOV-21G cells in relation to controls. Those proteins were enriched in 89 statistically significant signaling pathways and 62 statistically significant biological processes. These findings clearly demonstrated that SILAC quantitative proteomics was a useful and reliable method to study ivermectin-related proteomic changes in cancer cells, which in combination with molecular pathway networks and biological processes enrichments provided more comprehensive insights into molecular mechanisms of ivermectin in inhibiting TOV-21G cells.

Keywords: stable isotope labeling with amino acids in cell culture (SILAC), ovarian cancer, ivermectin, quantitative proteomics, bioinformatics analysis, molecular pathway, network, biological processes, biomarkers

1. Introduction

Stable isotope labeling with amino acids in cell culture (SILAC) is a polypeptide-labeling technology developed by the Thermo Fisher company of the United States in 2002 [1]. Heavy isotopes (^{13}C or ^{15}N) and light isotopes (^{12}C or ^{14}N) are used to label two essential amino acids (L-lysine and L-arginine) that are contained in a cell-cultured medium, respectively. After the cells were cultured with essential amino acids for 6–10 generations, all proteins were labeled with heavy isotopes or light isotopes. The cellular proteins stimulated by different treatment factors are analyzed by mass spectrometry (MS) to obtain the qualitative and quantitative proteome data [2]. SILAC generally allows heavy and light isotopes-labeled sample cells at the early stage of the experimental workflow, so the variability caused by the sample handling process was minimized [3]. SILAC was widely used in quantitative proteomics to study pathogenesis, drug target, protein modification and dynamics, protein-molecule interaction, and screen special functional proteins [4]. SILAC showed outstanding performance for quantification and dynamics of phosphosites in colorectal cancer with the treatment of the epidermal growth factor receptor (EGFR)-blocking antibody cetuximab, rendering it the effective method for cellular signaling study in cell culture models [5]. In terms of identification of protein-molecule interaction, SILAC combined with various affinity purifications of protein experimental setups could be used to distinguish specific complexes from nonspecific ones [6]. One study performed SILAC to overcome the most challenging problem in defining specific partners in protein complexes. The cells containing an affinity tagged protein were cultured in a light isotopic medium, while wild-type cells were grown in a heavy isotopic medium. The results of MS showed that specific partners appear as isotopically light [7]. SILAC also offers numerous opportunities to discover potential biomarkers and therapeutic targets for some drugs [8]. SILAC in combination with other developed approaches made SILAC more popular; for example, these SILAC labels in pulse or pulse-chase scenarios could be used to measure macromolecular dynamics on time scales of several hours [9]. An MS-based approach combining dynamic-SILAC labeling with isobaric mass tagging was well used to understand protein degradation and synthesis in cellular systems [10]. SILAC provided an effective scheme to comprehensively and systematically qualify and quantify complex mammalian cell proteome, which would promote progress in the medical field.

Ivermectin, marketed in 1981, was commonly used as a broad-spectrum anti-parasitic compound. It was approved to treat onchocerciasis (150–200 $\mu g\ kg^{-1}$ body weight), scabies (200 $\mu g\ kg^{-1}$ body weight), lymphatic filariasis (150–200 $\mu g\ kg^{-1}$ body weight), demodicosis (200 $\mu g\ kg^{-1}$ body weight), strongyloidiasis (200 $\mu g\ kg^{-1}$ body weight), pediculosis (400 $\mu g\ kg^{-1}$ body weight), and filariasis (due to Mansonella ozzardi, 6 mg as a single dose) [11].

Because ivermectin mainly targets chloride-dependent channels (γ-aminobutyric acid and glutamate), its safety could be fine in higher animals. In humans, especially the blood-brain barrier can reduce ivermectin delivery to the central nervous system [12]. The safety of ivermectin has been proved with clinical studies on children, infants, and pregnant women. A study including 170 infants and children with the treatment of oral ivermectin (mean dose = 223 $\mu g\ kg^{-1}$) showed good tolerance, and only seven subjects occurred mild adverse events [13]. A study including 893 pregnant women with the oral treatment of ivermectin also showed good tolerance, and no patients were reported to generate serious events (stillbirths, neonatal death, low birth weight, spontaneous abortions, preterm births, and congenital anomalies) [14]. Those studies proved that ivermectin was safe enough to be used in human diseases, but there was still insufficient evidence to prove no adverse side effects. The highest ivermectin dose was 200 μg/kg, which was approved by FDA. However, some patients without serious events have used 10 times more than the FDA-approved dose [15]. All those made ivermectin more likely to achieve success in clinical application. In recent days, studies found that ivermectin was effective in a completely new range of diseases, such as neurological disorders, antiviral (e.g., dengue, HIV, and encephalitis), antibacterial (e.g., Buruli ulcer and tuberculosis), anticancer (melanoma, lymphoid leukemia, lung cancer, glioblastoma, and breast cancer) [11]. The functions and mechanisms of ivermectin on anticancer generated interest and excitement in the scientific community. Ivermectin suppresses breast cancer by disrupting cellular signaling in the process and activating cytostatic autophagy through mediating PAK1 expression [16]. Ivermectin showed a synergistic effect with the chemotherapy agents by increasing cell death in leukemia cells. Some researchers, who aimed at overcoming cancer, claimed that ivermectin could be rapidly advanced into clinical trials [17]. Further study on molecular network, signaling pathway, and key biological processes of ivermectin would provide more useful information about this multifaceted "wonder" drug.

This chapter describes that SILAC identifies differentially expressed proteins (DEGs) in ivermectin-treated ovarian cancer cells in the following aspects: (i) ovarian cancer cell culture—TOV21G and labeled with heavy and light SILAC reagents; (ii) ivermectin treatment of SILAC-labeled TOV21G cells and protein preparation; (iii) the quality of SILAC-labeled protein samples with 1D SDS-PAGE; (iv) trypsin-digestion of SILAC-labeled proteins; (v) each fraction was subjected to LC-MS/MS analysis; and (vi) bioinformatics analysis (signaling pathway and biological process). SILAC can be a useful and effective method to detect protein alterations and dynamic changes in living cells, and the results would provide scientific data to further clarify molecular mechanisms of ivermectin in ovarian cancer.

2. Methods

2.1 Ovarian cancer cell culture

The human ovarian cell line (TOV-21G) was cultured in RPMI 1640 medium supplemented with 10% fetal bovine serum (FBS) and 5% CO_2 at 37°C. (i) Ovarian cell line used here was TOV-21G, which was obtained from Keibai Academy of Science (Nanjing, China) [18]. (ii) RPMI 1640 was used without glutamine, lysine, and arginine. (iii) FBS was brought from Gibco® Certified Thermo Fisher Scientific. (iv) The growing states of TOV-21G were observed, and the medium was changed in every 2 days.

2.2 SILAC medium preparation and labeling

SILAC "light" or "heavy" labeling growing medium (Thermo Fisher Scientific, US) was used to culture TOV-21G cells for 10 passages, to ensure a high level of stable isotope replacing original amino acids [8]. (i) "Light" labeled amino acids: 50 mg L-arginine HCl (Arg0), 50 mg L-lysine HCl (Lys0). "Heavy" labeled amino acids: 50 mg L-arginine-$^{13}C6,^{15}N4$ HCl (Arg10), 50 mg L-lysine-$^{13}C6,^{15}N2$ HCl (Lys8). (ii) A total of 1 L RPMI 1640 without glutamine, lysine, and arginine. (iii) For SILAC experiments, 50 mg Arg0 and 50 mg Lys0 ("light" labeling reagent) were added into 500 mL RPMI 1640 medium to form SILAC "light" labeling growing medium. A total of 50 mg Arg10 and 50 mg Lys8 ("heavy" labeling reagent) were added to 500 mL RPMI 1640 medium to form SILAC "heavy" labeling growing medium. (iv) The growing states of TOV-21G were observed, and the medium was changed in every 2 days. TOV-21G cells were cultured and passaged for 10 generations in 10-cm culture flasks.

2.3 Ivermectin treatment of SILAC-labeled TOV21G cells

When SILAC-labeled TOV21G cells achieved 80% cell density in 10-cm culture flasks, 20 μM ivermectin was added to TOV-21G cells in SILAC "heavy" labeling growing medium, and 0.1% DMSO was added to TOV-21G cells in SILAC "light" labeling growing medium for 24 h [19]. (i) Ivermectin ($C_{48}H_{74}O_{14}$, purity $\geq$ 95%): The drug was brought from Solarbio (http://www.solarbio.com/goods-3911.html). (ii) TOV-21G cells were counted, and 8000 cells/well were seeded in 96-well plates. Ivermectin was added into each well in different drug concentrations (0 μM, 10 μM, 20 μM, 30 μM, 40 μM, and 50 μM) for 24 h. CCK8 (10 μL) was added into each well for 1 h to measure absorbance values (OD) at a wavelength of 450 nm. Lethal concentration 50 (IC50) was calculated according to OD values of each well in different ivermectin concentrations. (iii) Ivermectin treatment group: 20 μM ivermectin was added to TOV-21G cells in SILAC "heavy" labeling growing medium for 24 h. Control group: 0.1% DMSO was added to TOV-21G cells in SILAC "light" labeling growing medium for 24 h.

2.4 Cell lysis and protein preparation

Ivermectin treatment and 0.1% DMSO treatment TOV-21G cells were collected and lysed by protein isolation buffer, respectively. (i) TOV-21G cells collection: A total of 500 μL trypsin was added to each 10-cm culture flasks for several minutes and collected with centrifugation (800 × g, 5 min). TOV-21G cells were washed with ice-cold phosphate buffer solution (PBS) for three times. (ii) Protein isolation buffer: 2 mM thiourea, 4% CHAPS (3-[(3-cholamidopropyl)-dimethylammonio] -1-propane), 7 M urea, 100 mM dithiothreitol (DTT), and 2% ampholyte. (iii) TOV-21G cells lysis: A total of 200 μL protein isolation buffer was added to each 10-cm culture flasks for 30 min (ice-cold) and then oscillated with five vortex cycles. (iv) SILAC-labeled protein collection: Protein isolation buffer was centrifuged (13,000 × g, 20 min, 4°C), and the SILAC-labeled protein samples were collected from the supernatants in new tubes. (v) Protein concentrations measurement: Protein concentrations of the SILAC-labeled protein samples were measured with the 2-D quant protein assay kit (Bio-Rad, US).

2.5 The preliminary experiment of SILAC-labeled protein samples

The "heavy"- and "light"-SILAC-labeled proteins were mixed and loaded onto 1X SDS-PAGE to check the quality. SDS-PAGE-separated proteins were further analyzed with MS/MS as a preliminary experiment to check the labeling efficiency. (i) The loading sample preparation: According to the 1:1 ratio, the "heavy"- and "light"-SILAC-labeled proteins were mixed in a 5X loading buffer. (ii) Electrophoresis: The mixed SILAC-labeled proteins were loaded onto SDS-PAGE gel (gel concentration: 12.5%) with the amount of 20 μg/lane by constant current (14 mA, 90 min). (iii) Coomassie brilliant blue staining: Prepare Coomassie brilliant blue stain and destain solutions. Filter the stain solution through Whatman 1 filter paper. (iv) MS/MS: Proteins were separated from SDS-PAGE bands, and then were reduced, alkylated, and trypsin-digested. The tryptic peptides were analyzed with MS/MS.

2.6 Trypsin-digestion of SILAC-labeled proteins

The main reagents and methods included: (i) Reducing agent: 100 mM DTT was added to SILAC-labeled protein sample. (ii) Uranyl acetate (UA) buffer: The UA buffer contained 8 M urea and 0.1 M Tris/HCL. The SILAC-labeled protein sample with DTT was filtered by a 10-kD ultrafiltration centrifuge tube for two times. (iii) Isolation mixture reacted: A total of 100 μL of 0.05 M iodoacetamide was added to the isolation mixture following centrifugalization (14,000 × g, 15 min). A total of 25 mM ammonium bicarbonate (NH_4HCO_3) was added to the mixture following centrifugalization (14,000 × g, 15 min). (iv) Trypsin buffer: 2 μg trypsin in 40 μL 100-mM NH_4HCO_3. (v) Tryptic peptide content: A volume of trypsin buffer (40 μL) was added to the mixture from last step and shaken by 600 rpm for 1 min. Enzymatic hydrolysis of the mixture was done for 16–18 h at 37°C. A volume (40 μL) of 25 mM NH_4HCO_3 was added to the mixture and that mixture was centrifuged (14,000 × g, 15 min); the filtrate was collected.

2.7 LC-MS/MS analysis

The instrument and materials are as follows: (i) MS instrument, for example, Q Exactive mass spectrometer (Thermo Fisher Scientific); (ii) Easy nLC system, for example, Proxeon Biosystems (Thermo Fisher Scientific); (iii) Thermo scientific EASY column: Acclaim PepMap, 100 μm × 2 cm, nanoViper, 5 μm-C18; (iv) analytical column: Thermo scientific EASY column (75 μm * 100 mm 3 μm-C18); (v) solvent A: 0.1% formic acid in H_2O; and (vi) solvent B: 0.1% formic acid, 84% acetonitrile in H_2O.

2.8 Search protein database with MaxQuant software

The main parameters are following [20]: Main search ppm: 6; missed cleavage: 2; MS/MS tolerance ppm: 20; de-isotopic: TRUE; enzyme: trypsin/P; database: uniprot_Homo_sapiens_169753_20190313; fixed modification: carbamidomethyl (C); lables: Lys(8), Arg(10); variable modification: oxidation (M), acetyl (protein N-term) decoy database; pattern: reverse; peptide FDR: 0.01; and protein FDR: 0.01.

2.9 Bioinformatics analysis

Several bioinformatics analyses were used, which are as follows: (i) The enrichment of Kyoto Encyclopedia of Genes and Genomes (KEGG) pathway was performed with R package clusterProfiler-KEGG (https://bioconductor.org/packages/release/bioc/html/clusterProfiler.html). (ii) The enrichment of biological processes (BPs) was analyzed with Cytoscape ClueGO. (iii) The level of statistical significance was set as $p < 0.05$ and adjusted p value < 0.05. For KEGG and BPs enrichment analyses, a Benjamini-Hochberg multiple text was used to adjust p value.

3. Results and discussion

3.1 Proteomic profile based on SILAC in ivermectin-treated TOV-21G cells

The flow chart of SILAC quantitative proteomics was shown to summarize the overall analysis process for the identification of ivermectin-related proteins (**Figure 1**).

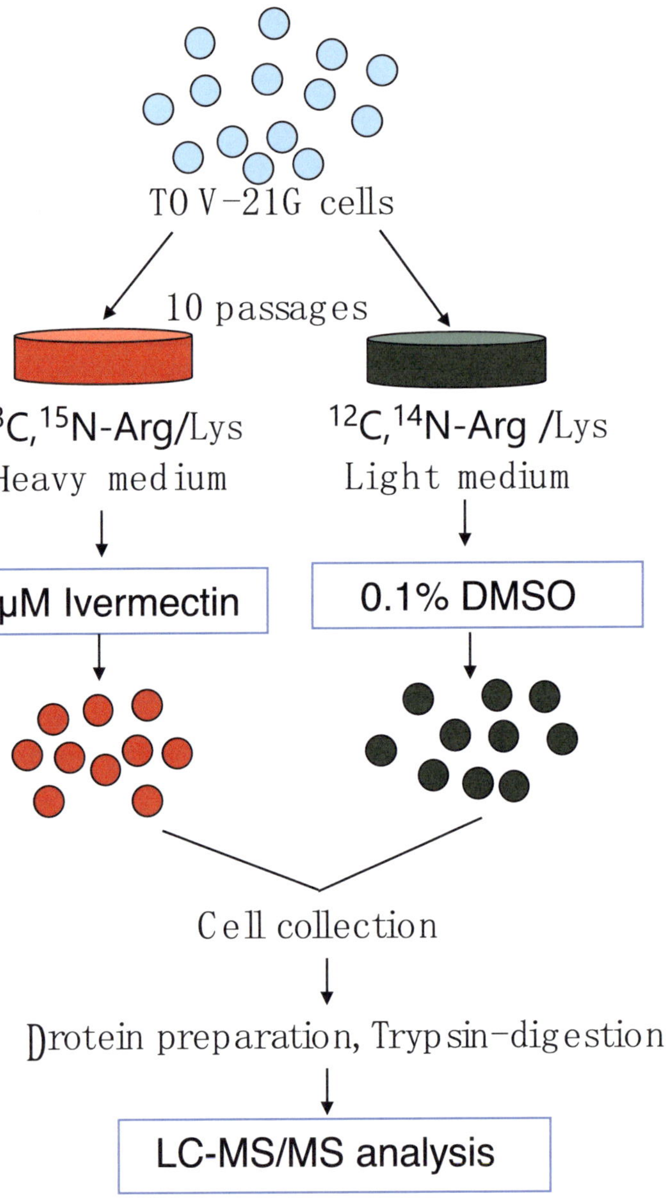

Figure 1.
The flow chart of SILAC quantitative proteomics analysis of ovarian cancer cells treated with and without ivermectin.

In total, 4447 proteins were identified with SILAC quantitative proteomics in human ovarian cancer cells treated with ivermectin. The ratio of "heavy"/"light" labeling samples was obtained, including 97.91% proteins with ratio < 1, and 2.09% proteins with ratio > 1. The MS/MS spectra of tryptic peptides EYQDLLNVK (**Figure 2A**) and VVQGSLDSLPQAVR (**Figure 2B**) are taken as examples. For peptide EYQDLLNVK (gene name = NEFM), the excellent b-ion and y-ion series were obtained with high signal-to-noise (S/N) (**Figure 2A**). For peptide VVQGSLDSLPQAVR (gene name = PKC1), the excellent b-ion and y-ion series were also obtained with high signal-to-noise (S/N) (**Figure 2B**).

The fold-changes of some identified proteins were very striking; for example, those upregulated proteins (ratio > 2), including histone H2A, progranulin,

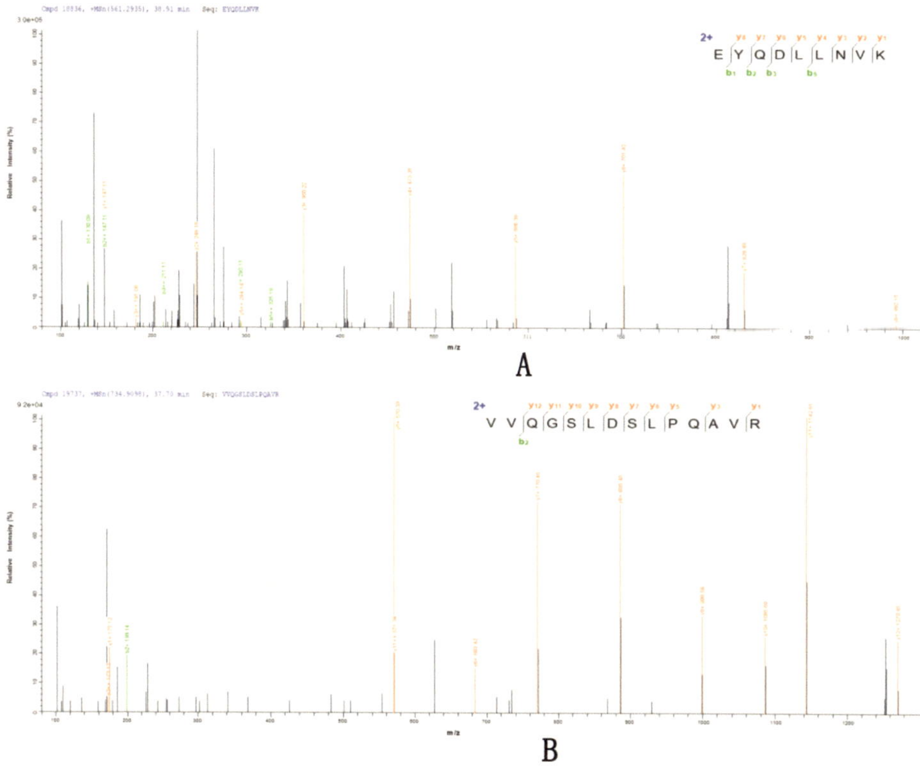

Figure 2.
The MS/MS spectra of tryptic peptides with SILAC labeling. (A) MS/MS spectrum of tryptic peptide EYQDLLNVK (gene name = NEFM). (B) MS/MS spectrum of tryptic peptide VVQGSLDSLPQAVR (gene name = PKC1).

cathepsin Z, SWI/SNF-related matrix-associated actin-dependent regulator of chromatin subfamily A-like protein 1, beta-mannosidase, GRAM domain-containing 1C, BMP-2-inducible protein kinase, ribosomal protein L3, ubiquitin-conjugating enzyme E2, PIK3R1, LIM domain-containing protein 1, retinal guanylyl cyclase 1, telomerase-binding protein EST1A, and COX7A2L protein. Some of them have been reported in ovarian cancers. For example, recent studies demonstrated that PI3K/AKT/mTOR and ERK1/2 signaling pathways were involved in this chemo-resistance. Progranulin was upregulated in epithelial ovarian cancer cell lines and associated with cisplatin resistance through regulating AKT/mTOR and ERK1/2 signaling pathways [21]. Progranulin is also involved in the process of cartilage development, progression, wound healing, and inflammation in ovarian cancer [22]. Additionally, one study showed that progranulin could directly activate cancer-associated fibroblasts to induce the epithelial-mesenchymal transition process of epithelial ovarian cancer cells [23]. Copy number loss of PIK3R1 most commonly occurs in ovarian cancer, which would activate AKT and p110-independent JAK2/STAT3 signaling and renders ovarian cancer cells vulnerable to AKT inhibitors [24]. CD97 can activate NF-κB-dependent JAK2/STAT3 pathway, consequently playing an important role in migratory, invasive capacity, and drug-resistant in ovarian cancer cells [25].

Some downregulated proteins (ratio < 0.1) were also very striking; for example, anion exchange protein, Rho guanine nucleotide exchange factor 16, dynein assembly factor 1, glucoside xylosyltransferase 1, glutamine synthetase, insulin-like growth factor-binding protein 2, microtubule-associated protein RP/EB family member 3, myelin proteolipid protein, neurochondrin, neurofilament medium polypeptide, phosphoenolpyruvate carboxykinase, ANKUB1, and TASOR 2. Some of them play an important role in the pathogenesis of ovarian cancer. For example, the most prominent effects of insulin-like growth factor-binding protein 2 in ovarian cancer include promoting driving invasion, proliferation, and suppressing apoptosis. The area under the ROC curve of insulin-like growth factor-binding protein 2 in detecting ovarian cancer was 0.815 (95% CI: 0.721–0.910, $P < 0.001$), further studies are needed to confirm its diagnostic performance at an early stage of ovarian cancer [26].

The glutamine metabolism could be a novel therapeutic target against cisplatin resistance in various cancers. Glutamine synthetase can take part in the reprogramming of glutamine metabolism to induce cisplatin resistance in A2780 ovarian cancer cells [27]. Anion exchanger 2 is a sodium-independent chloride/bicarbonate transporter, which is implicated in the regulation of membrane potential and intracellular potential of hydrogen (pH value). Anion exchanger 2 was highly expressed in ovarian cancer tissues compared to adjacent non-tumor lesions with quantitative proteomics analysis [28]. Those identified proteins in ovarian cancer cells treated with and without ivermectin with SILAC quantitative proteomics discovered reliable and effective biomarkers and drug targets for the anticancer process of ivermectin [8].

3.2 Ivermectin-mediated molecular pathway in human ovarian cancer cells

In total, 89 statistically significant molecular pathways were enriched based on those 4447 ivermectin-related proteins with KEGG pathway analysis (**Table 1**).

These molecular pathways demonstrated that ivermectin was involved in multiple cancer-related molecular pathways, such as mismatch repair process, ErbB signaling pathway, HIF-1 signaling pathway, cell-cycle regulation, ubiquitin-mediated proteolysis, AMPK signaling pathway, apoptosis, ferroptosis, proteoglycans, and central carbon metabolism in cancer. These molecular pathways also indicated that ivermectin was involved in multiple cancer pathogenesis, such as energy metabolism pathways, immunity-related pathways, stromal element-related pathways, RNA regulation pathways, hormone signaling pathways, and biosynthesis of substances. Different pathways enriched different proteins, whereas some pathways shared the same proteins. These data showed that ivermectin has a complex influence on various signaling pathways. The results were consistent with many studies previously. Ivermectin induced PAK1-mediated cytostatic autophagy both *in vitro* and *in vivo*, which might be one of the PAK1 inhibitors and inhibits the growth of ovarian cancer, glioblastoma, breast cancer, and NF2 tumors [29]. Another investigation found that ivermectin induced apoptosis in HeLa cells by upregulating Bax and p53 expressions, enhancing cytochrome c release, decreasing the levels of CDK2, CDK6, cyclin E, and cyclin D1 [30]. The primary immunogenic features—immunogenic cell death (ICD) included the release of high-mobility-group protein B1, secreted ATP, and surface-exposed calreticulin. Recent data supported that ivermectin could kill human triple-negative breast cancer cells through mechanisms of ICD, which induced pannexin-1 channel opening and cell death [31]. Ivermectin was also reported as an RNA helicase inhibitor, which reduced precursor and mature microRNAs potentially inhibiting cell invasion and proliferation [32]. Additionally, ivermectin effectively targets angiogenesis through decreasing membrane potential, mitochondrial respiration, ATP levels, and increasing mitochondrial superoxide, and the effects proliferation, capillary network formation, and survival in human brain microvascular endothelial cell [33]. According to the cross-talking between different signaling pathways, more favorable evidence indicated that ivermectin in combination with other drugs

Pathway ID	Pathway name	p value
hsa00010	Glycolysis/gluconeogenesis	2.03E−03
hsa00020	Citrate cycle (TCA cycle)	7.38E−11
hsa00030	Pentose phosphate pathway	3.51E−04
hsa00052	Galactose metabolism	6.04E−03
hsa00062	Fatty acid elongation	1.09E−02
hsa00071	Fatty acid degradation	1.39E−05
hsa00190	Oxidative phosphorylation	3.47E−12
hsa00230	Purine metabolism	1.08E−02
hsa00240	Pyrimidine metabolism	2.55E−03
hsa00270	Cysteine and methionine metabolism	8.93E−07
hsa00280	Valine, leucine and isoleucine degradation	2.06E−08
hsa00480	Glutathione metabolism	1.05E−03
hsa00510	N-Glycan biosynthesis	4.31E−03
hsa00513	Various types of N-glycan biosynthesis	5.15E−03
hsa00520	Amino sugar and nucleotide sugar metabolism	1.04E−07
hsa00620	Pyruvate metabolism	1.11E−07
hsa00630	Glyoxylate and dicarboxylate metabolism	4.09E−03
hsa00640	Propanoate metabolism	1.16E−05
hsa00920	Sulfur metabolism	4.45E−03
hsa01040	Biosynthesis of unsaturated fatty acids	1.09E−02
hsa01200	Carbon metabolism	3.96E−16
hsa01212	Fatty acid metabolism	1.74E−08
hsa01230	Biosynthesis of amino acids	3.09E−08
hsa03008	Ribosome biogenesis in eukaryotes	3.34E−04
hsa03010	Ribosome	7.07E−22
hsa03013	RNA transport	8.04E−20
hsa03015	mRNA surveillance pathway	4.28E−08
hsa03018	RNA degradation	2.24E−07
hsa03030	DNA replication	8.18E−09
hsa03040	Spliceosome	5.79E−24
hsa03050	Proteasome	1.50E−14
hsa03410	Base excision repair	1.22E−02
hsa03420	Nucleotide excision repair	4.72E−06
hsa03430	Mismatch repair	4.33E−04
hsa04012	ErbB signaling pathway	6.27E−03
hsa04066	HIF-1 signaling pathway	1.31E−05
hsa04071	Sphingolipid signaling pathway	9.01E−03
hsa04110	Cell cycle	4.03E−03
hsa04120	Ubiquitin mediated proteolysis	1.82E−06
hsa04130	SNARE interactions in vesicular transport	1.18E−04
hsa04137	Mitophagy—animal	9.07E−03

Pathway ID	Pathway name	p value
hsa04141	Protein processing in the endoplasmic reticulum	1.07E−13
hsa04142	Lysosome	3.81E−06
hsa04144	Endocytosis	9.24E−15
hsa04145	Phagosome	8.66E−07
hsa04152	AMPK signaling pathway	3.49E−03
hsa04210	Apoptosis	1.07E−03
hsa04213	Longevity regulating pathway—multiple species	4.46E−03
hsa04216	Ferroptosis	4.75E−04
hsa04510	Focal adhesion	2.57E−05
hsa04520	Adherens junction	2.04E−05
hsa04530	Tight junction	6.21E−05
hsa04540	Gap junction	3.14E−03
hsa04611	Platelet activation	4.03E−03
hsa04666	Fc gamma R-mediated phagocytosis	2.41E−03
hsa04714	Thermogenesis	1.08E−09
hsa04720	Long-term potentiation	7.22E−03
hsa04721	Synaptic vesicle cycle	2.29E−04
hsa04810	Regulation of actin cytoskeleton	4.94E−06
hsa04910	Insulin signaling pathway	2.90E−04
hsa04919	Thyroid hormone signaling pathway	5.20E−03
hsa04922	Glucagon signaling pathway	3.38E−04
hsa04931	Insulin resistance	1.26E−02
hsa04932	Nonalcoholic fatty liver disease (NAFLD)	1.44E−07
hsa04961	Endocrine and other factor-regulated calcium reabsorption	1.83E−03
hsa04962	Vasopressin-regulated water reabsorption	9.81E−03
hsa05010	Alzheimer disease	3.09E−07
hsa05012	Parkinson disease	5.32E−23
hsa05014	Amyotrophic lateral sclerosis (ALS)	2.55E−03
hsa05016	Huntington disease	7.93E−13
hsa05100	Bacterial invasion of epithelial cells	5.95E−10
hsa05110	*Vibrio cholerae* infection	6.22E−06
hsa05120	Epithelial cell signaling in *Helicobacter pylori* infection	1.07E−04
hsa05130	Pathogenic *Escherichia coli* infection	6.13E−09
hsa05131	Shigellosis	2.10E−07
hsa05132	Salmonella infection	7.10E−14
hsa05134	Legionellosis	1.05E−03
hsa05135	Yersinia infection	4.64E−06
hsa05163	Human cytomegalovirus infection	5.57E−05
hsa05165	Human papillomavirus infection	4.68E−03
hsa05169	Epstein-Barr virus infection	2.03E−05

Pathway ID	Pathway name	p value
hsa05170	Human immunodeficiency virus 1 infection	4.59E−08
hsa05203	Viral carcinogenesis	6.98E−05
hsa05205	Proteoglycans in cancer	2.44E−05
hsa05211	Renal cell carcinoma	2.65E−03
hsa05212	Pancreatic cancer	6.79E−03
hsa05220	Chronic myeloid leukemia	1.31E−02
hsa05230	Central carbon metabolism in cancer	1.18E−03
hsa03060	Protein export	8.73E−05

Table 1.
Statistically significant pathways identified with ivermectin-related proteins with KEGG pathway enrichment analysis.

exhibited more powerful anticancer effects, including daunorubicin, anti-BRAF V600 inhibitors, cytarabine, paclitaxel, and tamoxifen [34]. Obviously, the detailed mechanisms of ivermectin remain unclear. However, the application of new drugs brought ones to better health.

3.3 Ivermectin-mediated biological processes in human ovarian cancer cells

A total of 61 statistically significant biological processes were enriched based on those 4447 ivermectin-related proteins with GO analysis (**Figure 3**). These biological processes indicated that ivermectin was involved in multiple cancer-related biological processes, such as negative/positive regulation of canonical Wnt signaling pathway, cysteine-type endopeptidase activity involved in the apoptotic process, innate immune response-activating signal transduction, protein targeting to membrane, T-cell receptor signaling pathway, regulation of protein ubiquitination, activation of protein kinase activity, regulation of transcription by RNA polymerase II, and DNA-binding transcription factor activity. These results were consistent with many studies previously. For example, CTNNB1 (catenin beta 1, IMPβ1) in the biological process of protein polyubiquitination, was responsible for the nuclear entry of cargoes. Ivermectin can impact thermal stability and α-helicity of IMPα and IMPβ1 by binding to the IMPα armadillo repeat domain [35]. CASP3 in the biological process of protein kinase regulator activity is a member of the cysteine-aspartic acid protease (caspase) family. SK-MEL-28 cells were treated with different concentrations of ivermectin (2.5 μM, 5 μM, and 10 μM). Ivermectin enhanced the apoptosis effect by the upregulation of caspase-3 activity [36]. Also, PAK1 in the biological process of protein kinase regulator activity binds to and inhibits the activity of cyclin-cyclin-dependent kinase 2 or -cyclin-dependent kinase 4 complexes, and thus functions as a regulator of cell-cycle progression at G1. Ivermectin inhibited cancer stem cells formation by regulating the binding of PAK1/Stat3 complex and the IL-6 promoter [37]. YAP1 in the biological process of positive regulation of canonical Wnt signaling pathway was involved in the development, growth, repair, and homeostasis of multiple cancers. Ivermectin inhibited YAP1 nuclear expression and nuclear accumulation in gastric cancer cells. Moreover, in xenografts of gastric cancer cells, ivermectin suppressed tumor growth by regulating YAP1 nuclear expression [38]. Those identified proteins in ovarian cancer cells treated with and without ivermectin based on the SILAC method play important roles in multiple cellular signaling pathways and have broad biological activities. Those findings provide basic data for further study of ivermectin in ovarian cancer.

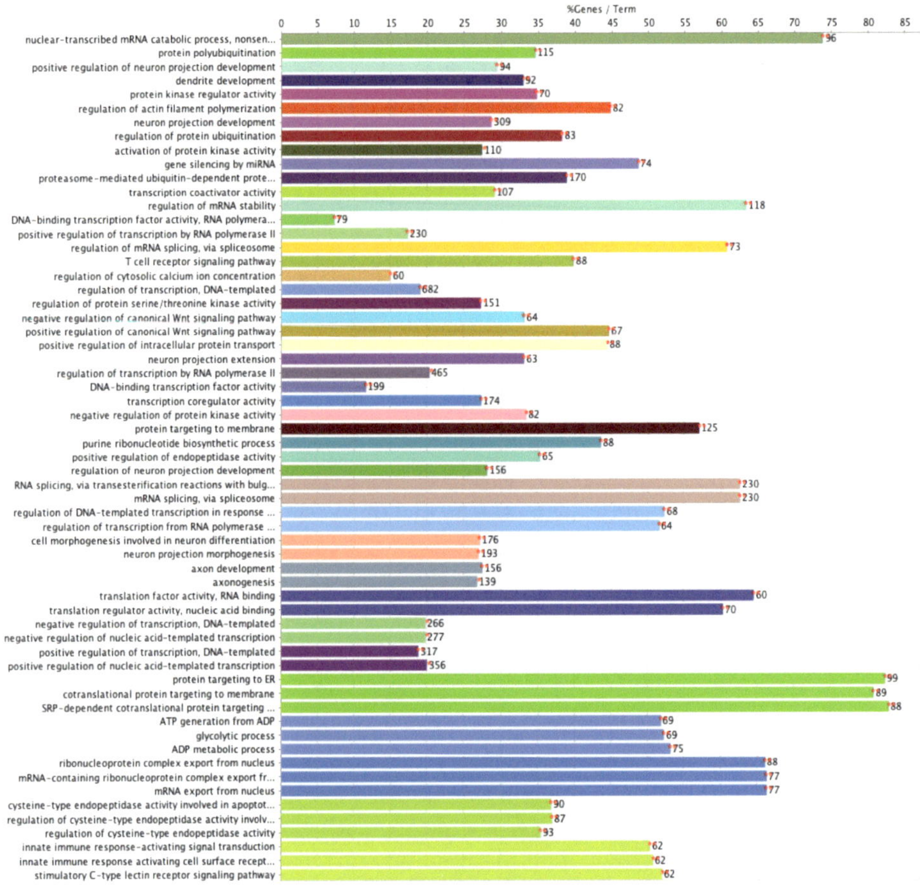

Figure 3.
Statistically significant biological processes (BPs) identified with ivermectin-related proteins with GO enrichment analysis.

4. Conclusions

Stable isotope labeling with amino acids in cell culture (SILAC) was an effective quantitative proteomics method to identify differentially expressed proteins or differentially modified proteins in cultured cells between two different conditions. In this study, ovarian cancer cells TOV-21 under two different conditions were cultured with the "heavy" labeling medium that contained 50 mg L-lysine-2HCl [^{13}C6, ^{15}N2] and 50 mg L-arginine-HCl [^{13}C6, ^{15}N4] in 500 mL RPMI 1640 medium, and the "light" labeling medium that contained 50 mg L-lysine-2HCl [^{12}C6, ^{14}N2] and 50 mg L-arginine-HCl [^{12}C6, ^{14}N4] in 500 mL RPMI 1640 medium for 10 passages, respectively. Then TOV-21G cells with SILAC "heavy" or "light" labeling were treated with or without 20 μM ivermectin for 24 h. The heavy- and light-stable isotope-labeled proteins were equally mixed (1:1), digested with trypsin, and analyzed with LC-MS/MS. A total of 4447 proteins were identified in ivermectin-treated TOV-21G cells relative to controls, and these proteins were significantly enriched in 89 molecular pathways, and 62 biological processes. These findings offer important data to study ivermectin-mediated molecular pathway network changes and discover effective ivermectin-related biomarkers and therapeutic targets for ivermectin treatment of ovarian cancer.

Acknowledgements

The authors acknowledge the financial supports from the Shandong First Medical University Talent Introduction Funds (to X.Z.), and the Hunan Provincial Hundred Talent Plan (to X.Z.).

Author's contributions

N.L. analyzed the data, prepared figures, and wrote the manuscript. X.Z. conceived the concept, designed the manuscript, wrote and critically revised the manuscript, and was responsible for the correspondence work and financial support.

Conflict of interest

We declare that we have no financial and personal relationships with other people or organizations.

Acronyms and abbreviations

BPs	biological processes
CTNNB1	catenin beta 1
ICD	immunogenic cell death
DTT	dithiothreitol
EGFR	epidermal growth factor receptor
IC50	lethal concentration 50
KEGG	Kyoto Encyclopedia of Genes and Genomes
LC	liquid chromatography
MS	mass spectrometry
MS/MS	tandem mass spectrometry
OD	absorbance values
PBS	phosphate buffer solution
SILAC	stable isotope labeling with amino acids in cell culture

Author details

Na Li[1,2] and Xianquan Zhan[1,2,3]*

1 Shandong Key Laboratory of Radiation Oncology, Shandong Cancer Hospital and Institute, Shandong First Medical University, Jinan, Shandong, P.R. China

2 Medical Science and Technology Innovation Center, Shandong First Medical University, Jinan, Shandong, P.R. China

3 Gastroenterology Research Institute and Clinical Center, Shandong First Medical University, Jinan, Shandong, P.R. China

*Address all correspondence to: yjzhan2011@gmail.com

References

[1] Chen X, Wei S, Ji Y, Guo X, Yang F. Quantitative proteomics using SILAC: Principles, applications, and developments. Proteomics. 2015;**15**(18): 3175-3192. DOI: 10.1002/pmic.2015 00108

[2] Deng J, Erdjument-Bromage H, Neubert TA. Quantitative comparison of proteomes using SILAC. Current Protocols in Protein Science. 2019;**95**(1):e74. DOI: 10.1002/cpps.74

[3] Hoedt E, Zhang G, Neubert TA. Stable isotope labeling by amino acids in cell culture (SILAC) for quantitative proteomics. Advances in Experimental Medicine and Biology. 2019;**1140**:531-539. DOI: 10.1007/978-3-030-15950-4_31

[4] Ong SE. The expanding field of SILAC. Analytical and Bioanalytical Chemistry. 2012;**404**(4):967-976. DOI: 10.1007/s00216-012-5998-3

[5] Stepath M, Zülch B, Maghnouj A, et al. Systematic comparison of label-free, SILAC, and TMT techniques to study early adaption toward inhibition of EGFR signaling in the colorectal cancer cell line DiFi. Journal of Proteome Research. 2020;**19**(2):926-937. DOI: 10.1021/acs.jproteome.9b00701

[6] Kaeser-Pebernard S, Diedrich B, Dengjel J. Identification and regulation of multimeric protein complexes in autophagy via SILAC-based mass spectrometry approaches. Methods in Molecular Biology. 2019;**1880**:341-357. DOI: 10.1007/978-1-4939-8873-0_23

[7] Tackett AJ, DeGrasse JA, Sekedat MD, et al. I-DIRT, a general method for distinguishing between specific and nonspecific protein interactions. Journal of Proteome Research. 2005;**4**(5):1752-1756. DOI: 10.1021/pr050225e

[8] Li N, Li J, Desiderio DM, Zhan X. SILAC quantitative proteomics analysis of ivermectin-related proteomic profiling and molecular network alterations in human ovarian cancer cells. Journal of Mass Spectrometry. 2021;**56**(1):e4659. DOI: 10.1002/jms.4659

[9] Snider J, Wang D, Bogenhagen DF, Haley JD. Pulse SILAC approaches to the measurement of cellular dynamics. Advances in Experimental Medicine and Biology. 2019;**1140**:575-583. DOI: 10.1007/978-3-030-15950-4_34

[10] Savitski MM, Zinn N, Faelth-Savitski M, et al. Multiplexed proteome dynamics profiling reveals mechanisms controlling protein homeostasis. Cell. 2018;**173**(1):260-274. e25. DOI: 10.1016/j.cell.2018.02.030

[11] Crump A. Ivermectin: Enigmatic multifaceted 'wonder' drug continues to surprise and exceed expectations. Journal of Antibiotics (Tokyo). 2017;**70**(5):495-505. DOI: 10.1038/ja.2017.11

[12] Develoux M. Ivermectin. Annales de Dermatologie et de Vénéréologie. 2004;**131**(6-7 Pt 1):561-570. DOI: 10.1016/s0151-9638(04)93668-x

[13] Levy M, Martin L, Bursztejn AC, et al. Ivermectin safety in infants and children under 15 kg treated for scabies: A multicentric observational study. The British Journal of Dermatology. 2020; **182**(4):1003-1006. DOI: 10.1111/bjd.18369

[14] Nicolas P, Maia MF, Bassat Q, et al. Safety of oral ivermectin during pregnancy: A systematic review and meta-analysis. The Lancet Global Health. 2020;**8**(1):e92-e100. DOI: 10.1016/s2214-109x(19)30453-x

[15] Guzzo CA, Furtek CI, Porras AG, et al. Safety, tolerability, and

pharmacokinetics of escalating high doses of ivermectin in healthy adult subjects. Journal of Clinical Pharmacology. 2002;**42**(10):1122-1133. DOI: 10.1177/009127002401382731

[16] Dou Q, Chen HN, Wang K, et al. Ivermectin Induces Cytostatic Autophagy by Blocking the PAK1/Akt Axis in Breast Cancer. Cancer Research. 2016;**76**(15):4457-4469. DOI: 10.1158/0008-5472.Can-15-2887

[17] Sharmeen S, Skrtic M, Sukhai MA, et al. The antiparasitic agent ivermectin induces chloride-dependent membrane hyperpolarization and cell death in leukemia cells. Blood. 2010;**116**(18): 3593-3603. DOI: 10.1182/blood-2010-01-262675

[18] Li N, Zhan X. Anti-parasite drug ivermectin can suppress ovarian cancer by regulating lncRNA-EIF4A3-mRNA axes. The EPMA Journal. 2020;**11**(2): 289-309. DOI: 10.1007/s13167-020-00209-y

[19] Li N, Li H, Wang Y, Cao L, Zhan X. Quantitative proteomics revealed energy metabolism pathway alterations in human epithelial ovarian carcinoma and their regulation by the antiparasite drug ivermectin: Data interpretation in the context of 3P medicine. The EPMA Journal. 2020;**11**(4):661-694. DOI: 10.1007/s13167-020-00224-z

[20] Trautwein-Schult A, Bartel J, Maaß S, Becher D. Metabolic labeling of clostridioides difficile proteins. Methods in Molecular Biology. 2021;**2228**:271-282. DOI: 10.1007/978-1-0716-1024-4_19

[21] Perez-Juarez CE, Arechavaleta-Velasco F, Zeferino-Toquero M, et al. Inhibition of PI3K/AKT/mTOR and MAPK signaling pathways decreases progranulin expression in ovarian clear cell carcinoma (OCCC) cell line: A potential biomarker for therapy response to signaling pathway inhibitors. Medical Oncology. 2019;**37**(1):4. DOI: 10.1007/s12032-019-1326-5

[22] Yang D, Wang LL, Dong TT, et al. Progranulin promotes colorectal cancer proliferation and angiogenesis through TNFR2/Akt and ERK signaling pathways. American Journal of Cancer Research. 2015;**5**(10):3085-3097

[23] Dong T, Yang D, Li R, et al. PGRN promotes migration and invasion of epithelial ovarian cancer cells through an epithelial mesenchymal transition program and the activation of cancer associated fibroblasts. Experimental and Molecular Pathology. 2016; **100**(1):17-25. DOI: 10.1016/j.yexmp.2015.11.021

[24] Li X, Mak VCY, Zhou Y, et al. Deregulated Gab2 phosphorylation mediates aberrant AKT and STAT3 signaling upon PIK3R1 loss in ovarian cancer. Nature Communications. 2019;**10**(1):716. DOI: 10.1038/s41467-019-08574-7

[25] Park GB, Kim D. MicroRNA-503-5p inhibits the CD97-mediated JAK2/STAT3 pathway in metastatic or paclitaxel-resistant ovarian cancer cells. Neoplasia. 2019;**21**(2):206-215. DOI: 10.1016/j.neo.2018.12.005

[26] Prayudi PKA, Budiana ING, Mahayasa PD, et al. Diagnostic accuracy of serum insulin-like growth factor-binding protein 2 for ovarian cancer. International Journal of Gynecological Cancer. 2020;**30**(11):1762-1767. DOI: 10.1136/ijgc-2020-001479

[27] Guo J, Satoh K, Tabata S, et al. Reprogramming of glutamine metabolism via glutamine synthetase silencing induces cisplatin resistance in A2780 ovarian cancer cells. BMC Cancer. 2021;**21**(1):174. DOI: 10.1186/s12885-021-07879-5

[28] Zhang LJ, Lu R, Song YN, et al. Knockdown of anion exchanger 2

suppressed the growth of ovarian cancer cells via mTOR/p70S6K1 signaling. Scientific Reports. 2017;**7**(1):6362. DOI: 10.1038/s41598-017-06472-w

[29] Wang K, Gao W, Dou Q, et al. Ivermectin induces PAK1-mediated cytostatic autophagy in breast cancer. Autophagy. 2016;**12**(12):2498-2499. DOI: 10.1080/15548627.2016.1231494

[30] Song D, Liang H, Qu B, et al. Ivermectin inhibits the growth of glioma cells by inducing cell cycle arrest and apoptosis in vitro and in vivo. Journal of Cellular Biochemistry. 2019;**120**(1):622-633. DOI: 10.1002/jcb.27420

[31] Draganov D, Gopalakrishna-Pillai S, Chen YR, et al. Modulation of P2X4/P2X7/pannexin-1 sensitivity to extracellular ATP via ivermectin induces a non-apoptotic and inflammatory form of cancer cell death. Scientific Reports. 2015;**5**:16222. DOI: 10.1038/srep16222

[32] Yin J, Park G, Lee JE, et al. DEAD-box RNA helicase DDX23 modulates glioma malignancy via elevating miR-21 biogenesis. Brain. 2015;**138**(Pt 9):2553-2570. DOI: 10.1093/brain/awv167

[33] Liu Y, Fang S, Sun Q, Liu B. Anthelmintic drug ivermectin inhibits angiogenesis, growth and survival of glioblastoma through inducing mitochondrial dysfunction and oxidative stress. Biochemical and Biophysical Research Communications. 2016;**480**(3):415-421. DOI: 10.1016/j.bbrc.2016.10.064

[34] Liu J, Zhang K, Cheng L, Zhu H, Xu T. Progress in Understanding the Molecular Mechanisms Underlying the Antitumour Effects of Ivermectin. Drug Design, Development and Therapy. 2020;**14**:285-296. DOI: 10.2147/dddt.S237393

[35] Yang SNY, Atkinson SC, Wang C, et al. The broad spectrum antiviral ivermectin targets the host nuclear transport importin α/β1 heterodimer. Antiviral Research. 2020;**177**:104760. DOI: 10.1016/j.antiviral.2020.104760

[36] Deng F, Xu Q, Long J, Xie H. Suppressing ROS-TFE3-dependent autophagy enhances ivermectin-induced apoptosis in human melanoma cells. Journal of Cellular Biochemistry. 2018. DOI: 10.1002/jcb.27490

[37] Kim JH, Choi HS, Kim SL, Lee DS. The PAK1-Stat3 signaling pathway activates IL-6 gene transcription and human breast cancer stem cell formation. Cancers (Basel). 2019;**11**(10):1527. DOI: 10.3390/cancers11101527

[38] Nambara S, Masuda T, Nishio M, et al. Antitumor effects of the antiparasitic agent ivermectin via inhibition of Yes-associated protein 1 expression in gastric cancer. Oncotarget. 2017;**8**(64):107666-107677. DOI: 10.18632/oncotarget.22587

Chapter 6

Nanotechnology Application and Intellectual Property Right Prospects of Mammalian Cell Culture

Harikrishnareddy Rachamalla, Anubhab Mukherjee and Manash K. Paul

Abstract

The significant challenges faced by modern-day medicine include designing a target-specific drug delivery system with a controlled release mechanism, having the potential to avoid opsonization and reduce bio-toxicity. Nanoparticles are materials with nanoscale dimensions and maybe natural and synthetic in origin. Engineered nano-sized materials are playing an indispensable role in the field of nanomedicine and nanobiotechnology. Besides, engineered nano-sized particles impart therapeutic applications with enhanced specificity because of their unique bespoke properties. Moreover, such application-customized nanoparticles offer an enormous possibility for their compatibility with different biological molecules like proteins, genetic materials, cell membranes, and organelles at the nano-bio frame. Besides, surface functionalization with targeting moieties such as small molecule ligands, monoclonal antibodies, aptamers, cell-penetrating peptides, and proteins facilitate nanoparticle-based specific tissue targeting. This review summarizes some of the advances in nanoparticle-based therapeutics and theranostics. A better understanding of idealistic preparation methods, physicochemical attributes, surface functionalization, biocompatibility can empower the potential translation of nanomaterials from the 'bench-to-bedside'. In modern-day medicine, engineered nanoparticles have a wide range of demands ranging from bio-imaging, theranostics, tissue engineering, sensors, drug and nucleic acid delivery, and other pharmaceuticals applications. 2D and 3D mammalian cell-based assays are widely used to model diseases, screening of drugs, drug discovery, and toxicity analyses. Recent advances in cell culture technology and associated progress in nanotechnology have enabled researchers to study a wide variety of physiologically relevant questions. This chapter explores the properties of nanoparticles, different targeted delivery methods, biological analysis, and theranostics. Moreover, this chapter also emphasizes biosafety and bioethics associated with mammalian cell culture and discusses the significance of intellectual property rights from an industrial and academic perspective.

Keywords: nanotechnology, intellectual property right, mammalian cell culture, nanoparticle biocompatibility, targeted drug delivery, bioethics

1. Introduction

Nanomaterials (NMs) are engineered chemical substances or materials with a particle size of 1–100 nm in diameter. Today NMs are extensively explored and engaged for commercial purposes in different fields, and many sophisticated NMs have shown great promise in biotechnology and biomedicine [1]. NMs display inimitable physicochemical attributes due to their size range in nanometers, high surface area, tunable surface charge, unique composition, various morphologies, and surface composition. Due to their remarkable physicochemical attributes, NMs are significantly different from their bulk materials of a similar symphony, allowing them to perform remarkably well with improved functionality, sensitivity, competence, and selectivity towards developing biomedicines. Various NMs are evaluated to get desired biomedical efficacy for nanomedicine-related applications, including different metal nanoparticles, liposomes, quantum dots, polymeric micelles, dendrimers, and carbon-based nanoparticles. Two critical mechanisms for delivering drug-loaded NMs to the diseased sites are passive targeting and active targeting. A passive targeting mechanism happens via enhanced permeability and retention (EPR) [2]. Inactive targeting mechanism relies on surface functionalized NMs with various biomarkers that bind with receptors over-expressed at the pathological tissue [3].

The importance of cell culture advances in the medical sector has long been recognized. Mammalian cell culture (MCC) entails first isolating cells from a specific organ tissue and then creating a culture in a suitable artificial setting. Disaggregation using different methods may be used to obtain preliminary separation of cells from the identified organ tissues. The isolated primary cells are typically obtained from an *in vivo* setting, although some cells come as established cell lines. MCCs are widely used in the biomedical field to investigate numerous applications [4]. Since cell culture-based studies provide highly stable and repeatable results, researchers consider this technique as an essential model system in cellular and molecular biology. MCC needs an ideal environment for development, which can be divided into nutritional and physicochemical requirements.

Nutritional necessities comprise an adherent substrate or growing medium that offers conditions like essential amino acids, sugars, vitamins, minerals, growth factors, hormones, and gases (O_2, CO_2). All these features regulate physicochemical factors such as pH, osmotic pressure, and temperature. Many cell lines need solid or semi-solid support in the form of a substrate, while others can be grown in a suspension culture medium. These technologies have evolved as a means of assessing the efficacy and side effects of novel active pharmaceutical ingredients (APIs), immunotherapeutic, and biopharmaceuticals [5]. Animal, plant, and bacterial cells are regularly cultured in fixed culture medium under precise laboratory circumstances; among this, animal-based cell cultures are more complex than others due to their genetic complexity. Directed differentiation of adult stem cells and pluripotent stem cell culture is another challenging aspect. Recent advances in stem cell culture technology have provided significant input for the successful culture of tissue-mimicking 3D organoids [4, 6].

In recent years, nanotechnology (NT) and associated disciplines have gained rapid escalation in biomedical implementations such as diagnosis, testing, tracking, drug delivery, nanomedicine, medical implants, and electronics due to their camaraderie with biological entities. Biomedicine embraces the design and synthesis of NMs, along with other nanoparticles (NPs) and nano-devices [7]. Once properly formulated, NMs show their natural aptitude to traverse with the blood flow via various routes based on their attributes and eventually get access to all the organs. Due to their intrinsic biocompatible interactions, the NPs exhibit unique physicochemical attributes associated with lesser immunogenicity and non-toxicity. There are numerous advantages of using NMs for various biological applications: i) it increases the concentration of drug in the pathological tissues and control the slow release of the drug; ii) it solves issues connected to the low solubility and bioavailability of the drug; and iii) enhanced biodegradability and biocompatibility iv) drugs/genes/imaging agents can be easily loaded due to their tunable surface functionalities [1, 7, 8]. Imaging agents could endow *in vivo* drug tracking ability to determine drug delivery efficacy during treatment. In recent years, various nanoparticles such as liposomes, polymers, metal nanoparticles, inorganic nanoparticles have been developed for selectively targeting tumor cells and other pathological tissues without causing any destruction to healthy cells or organs. In this chapter, the application of nanotechnology and Intellectual property rights (IPR) prospects of mammalian cell culture will be discussed in the subsequent sections.

2. Compatibility of nanomaterials towards biological interactions

NMs attract considerable interest due to their unique, tunable, versatile physicochemical properties, easy preparation methods, biocompatibility, and surface functionalization [1]. Nonetheless, the compatibility of the nanoparticles with biological entities constitutes the most fundamental phenomenon and highlights the importance of basic research [9]. Most bio-applications, including drug delivery, bioimaging, and treatment, start from the attachment of nanoparticles onto the target cells. The biocompatibility of nanoparticles depends on the physical and chemical attributes like diameter, shape, composition, concentration, functionalized moieties, and surface potential (**Figure 1**) [10]. Among the various NMs, Quantum dots have risen as an innovative bio-imaging tool due to their unique tunable physicochemical attributes. Existing research has guided the development of versatile quantum dots that are highly fluorescent and stable under diverse biological circumstances. Moreover, quantum dots with enclosed amphipathic polymers have been developed and surface-functionalized with receptor targeting ligands for bio-imaging and drug-delivery in animal models. Fascinatingly, these materials were found to be compatible with the cells. However, their complete chronic *in vivo* genotoxicity, blood, and organ compatibility need to be assessed [1, 7, 11].

Polymeric nanoparticles have drawn considerable attention in drug and gene delivery, tissue engineering, and many biomedical applications due to their non-toxic nature and high compatibility to biological systems. They are colloidal in nature and composed of natural or synthetic, or semi-synthetic polymers. In this perspective, biodegradable nanoparticles of the highly compatible triblock copolymer are used for non-viral gene transfections [3].

Liposomes are another popular nanomaterial drug delivery system that is best documented and adapted owing to their bio-congenial physicochemical properties [12]. Liposomes consist of unilamellar/ multilamellar lipid bilayers having an aqueous core inside. The nanoscale carrier system offers substantial advantages such as biodegradability, biocompatibility, ease of synthesis, less toxicity, sustained drug release, and the ability to incorporate hydrophilic and hydrophobic drugs. Liposomal-surface modification is a crucial strategy for targeted therapy and especially for cancer treatment [13]. Seventeen liposomal formulations are clinically approved for cancer, inflammation, infectious diseases, antibiotic drugs, and anesthetics, while several liposomal formulations are under various phases of clinical trials [12, 13].

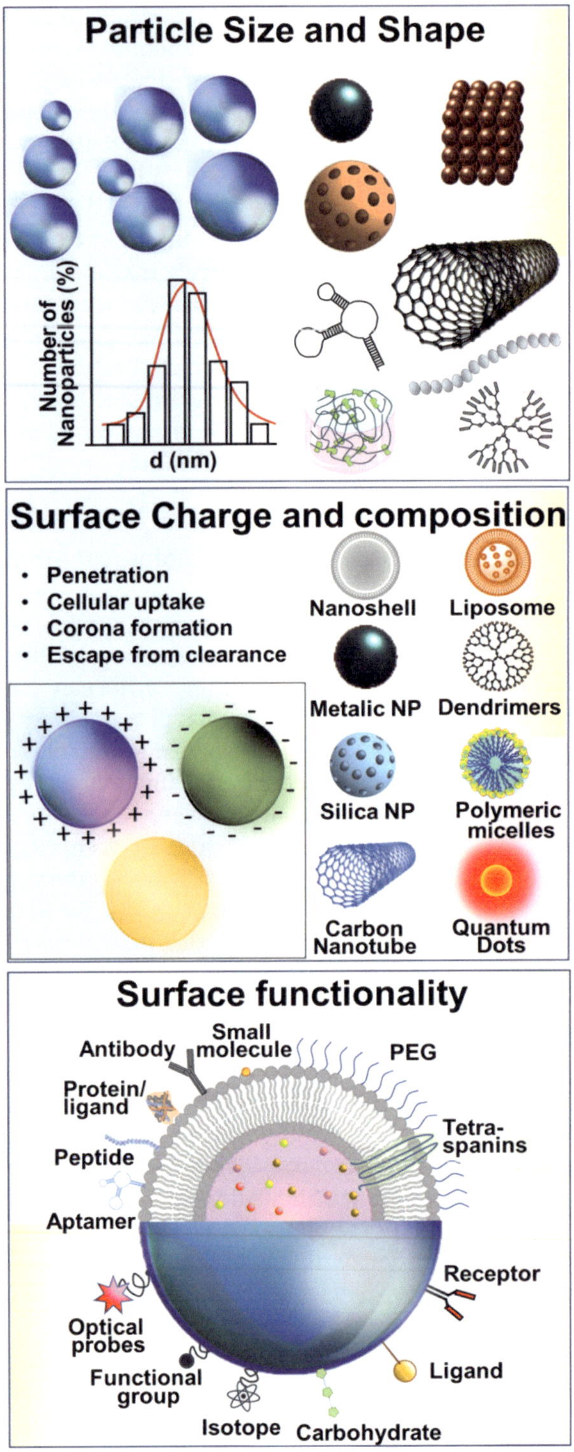

Figure 1.
Precision of targeted drug delivery using nanocarriers and bio-compatibility. Nanoparticle based drug delivery platform depends on surface functionality, size and shape and surface charge and composition.

Despite several encouraging biomedical implementations of nanoparticles, the biocompatible assessments including, complete acute and chronic toxicological evaluation of NMs, are inadequately comprehended. Additionally, the toxicity of the nanoparticle design aims to find out favorable physicochemical attributes of different materials. Hence, the active bio-molecule with biological entities must be highly allied to the nanoparticles approaching direct contact with biological objects rather than its transient initial distribution. Much to our intrigue, various nanoparticles - liposomes, lipoplexes, polymeric nanoparticles, polyplexes, metal nanoparticles, metal oxides, dendrimers, and quantum dots are wisely engineered for their medical application like diagnosis, drug and gene delivery, tissue engineering, and biosensing [8, 13, 14]. Moreover, it is unavoidable to thoroughly assess and investigate compatibility/unwanted toxicity with nanoparticles to bring clinical success. The subsequent section will relate to how the physicochemical properties of engineered nanoparticles can be persuaded towards accomplishing the desired biological aspiration lacking any toxicological impact.

2.1 Tunable physicochemical attributes of nanomaterials compatible with biomedical applications

Nanoparticles exhibit outstanding physicochemical attributes which can be manipulated to harness the best possible benefits out of them - their tunable diameter, high surface area, various morphologies, different concentrations and compositions, surface functionalization, etc. [15] (**Figure 1**). Interactions of NMs to the cell surface, their internalization, and subcellular localization, communication with the cells eventually contribute to therapeutic or adverse effects. Understanding the physicochemical attributes of NMs and their interactions with biological entities can help design superior NMs for further applications. We are jotting down the relevant physicochemical attributes of nanoparticles, which may modulate their function in therapeutic or toxicity aspects; thus, they need to be engineered wisely [16].

2.1.1 Nanomaterials size

For engineered nanoparticles, the primary crucial feature is their dimensions/size, which partially governs other physicochemical characteristics. The reduced diameter of the nanoparticles, provide possibilities for high cellular localization making them interact with cellular tissues, especially pathological tissue to a greater extent to attain specific biological outcome for the remedial purpose. Size-dependent bio-distribution studies were performed using three different sizes containing (20, 50, and 200 nm) drug conjugated silica nanoparticles. It revealed that nanoparticles having 50 nm diameter had the highest tumor localization, enhanced cancer tissue retention, and slower clearance [16]. Moreover, nano-sized particles preside over their pharmacokinetics, are predictable to traverse biological barriers, which is not possible for bulk particles. Besides, ~50 nm diameter particles showed higher efficacy because of active engagement to the biological tissues, modulating pathways, and cellular activities [17].

2.1.2 Nanomaterials surface charge

The surface charge is a unique character of NMs to manage its therapeutic and toxicological effects and plays a significant role in electrostatic interactions of NMs and living entities (**Figure 1**) [10]. Besides, the cellular localization pathways and tissue interactions are regulated by the surface charge of the nanoparticles, thus playing a significant role in the compatibility and cellular toxicity. Several reports suggest that nanoparticles with a positive charge highly interact with the negatively charged cell membranes and provoke genotoxicity [18]. Positively charged cationic liposomal drug and gene delivery systems have been extensively studied for the last decade. It was recently shown that cationic lipoplexes are not showing any genotoxicological aberrations in the Swiss albino mice. Typically, cell membranes are anionic in nature; thus, negatively charged NMs have very slow cellular internalization compared to neutral and positive nanoparticles [14]. Surface potentials of metal particles in regulating different tumorous and non-tumorous tissue types are also established. Several studies have suggested the role of the surface potential of different nanoparticles and their interactions with the biological entities and how surface charge modulates their biological functions, which shed light to design and engineer nanoparticles for a selective cellular target for various diseases with minimal toxicity [16, 18].

2.1.3 Surface functionalization

Nanoparticles play a vital role in promoting intracellular delivery of encapsulated therapeutic agents and increase their retention in pathological tissues compared to healthy tissues [1]. Surface functionalization with suitable receptor-targeted ligands using different methods results in the formation of targeted nanoparticles with improved therapeutic response and minimized off-target side effects by prolonging their circulation time in blood, increasing target specificity, cellular uptake, and drug accumulation in the tumors (escaping lysosomal degradation and enhancing stimuli-responsive drug release) (**Figure 1**) [17, 19]. Depending on their application, nanoparticles are functionalized with different targeting ligands either by directly conjugating ligands to PEGylated nanoparticles through post-insertion technique or by covalent grafting on the surface of the nanoparticles. In this context, surface functionalization of nanoparticles with antibodies, peptides, folic acid, aptamers has been extensively studied. This prompts scientists to design and engineer nanoparticles for selective targeting and high retention in the tumor tissue rendering minimal toxicity to the vital organs [19, 20].

3. Mechanism of targeted drug delivery using nano-carrier

Nanoparticles play a vital role in promoting intracellular delivery of enclosed therapeutic agents and increase their retention in the different pathological tissues compared to other therapies [21]. Like normal tissues, tumors need nourishments by means of food and oxygen and a capacity to remove metabolic excretes and carbon dioxide. Diverse patterns of tumor-associated neovascularization, obtained by angiogenesis, cope with these demands. Primary conservative treatment modalities involved in cancer treatment are surgery, radiotherapy, and chemotherapy, while additional therapies such as immune therapy, targeted therapy, and hormone therapy are chosen depending on the type of tumor [22]. On the other hand, the failure of chemotherapeutic drugs to specifically target cancer tissue hinders many treatment modalities. It is habitually faster and economically cheaper to design an existing drug to encapsulate in a delivery system a more effective way to superior targeting of tissues than to invent a completely new one. The drug delivery mechanism can be classified into passive and active, respectively.

3.1 Nanoparticle drug delivery by passive targeting

Passive targeted drug delivery mainly depends on the physicochemical attributes of the NMs, such as shape, diameter, surface potentials, and pathophysiological conditions of the disease microenvironment. Intravenously injected drug encapsulated NMs tend to disperse throughout the body evenly [23]. However, unlike normal tissues, tumor cells tend to take up particles of a definite diameter to a greater extent than healthy cells due to the arrangement of capillary endothelial cells, accumulating extravasated molecules in the interstitial spaces poor lymphatic drainage increases the permeation and accumulation of drug-mediated NMs. This type of NMs accumulation in the tumor region is known as the EPR effect [1, 2]. The EPR effect is influenced by physicochemical attributes of NM including particle diameter, shape, and surface potentials greatly influence the circulation time, penetration speed, tumor localization, and intracellular internalization.

Particle diameter plays a critical role in achieving effective drug delivery as it enhances permeation and circulation time and reduces renal clearance. For example, phagocyte cells facilitate larger particle uptake, while non-phagocytic cells favor the uptake of smaller particles. PEGylated NPs reduced plasma protein adsorption on their surface and reduced hepatic filtration when their size is smaller than 100 nm [24]. Particle diameter with 20–200 nm effectively enhances the permeation in both hyper-permeable and poorly permeable tumors, and particles with less than 6 nm avoid renal clearance. The NPs surface potentials could play a vital role in circulation and cellular localization [24, 25]. NPs with positive surface potentials such as cationic liposomes induce non-specific interactions with blood components and aggregation of liposomes results in a reduction of EPR effect and increased renal clearance. However, positively charged NPs are more readily taken up by cancer cells. Whereas anionic and neutral surface potential-bearing NPs circulate longer in the blood circulation [1, 2, 24].

Besides, Polyethylene glycol (PEG) polymer is used as a stabilizer (stealth liposomes) that increases the circulation time in blood up to 24–48 hours and improves *in vivo* stability [26]. PEG-coated liposomes induce the 'steric stabilization effect' by creating hydrophilicity on the surface of liposomes that shield surface charge and increases the repulsive forces between liposomes and blood components. Thus, it prevents aggregation of liposomes and opsonization by the reticuloendothelial system, macrophages, mononuclear phagocytic cells and prolongs their systemic circulation. On the other hand, PEG-coated liposomes induce PEG-specific IgM antibodies, enhancing hepatic uptake and rapid clearance of liposomes from systemic circulation on subsequent administration. PEG corona produces steric hindrance with tumor cells that prevent effective internalization, which could be minimized by using short PEG chains with molecular weight less than 1000 Da or by designing PEG with enzyme-cleavable bound or tumor-targeting ligands [20, 26]. To investigate the influence of shape on the cellular localization of NPs, Li et al. conducted large-scale molecular simulations to evaluate different NP geometries with identical surface area, ligand-receptor interaction strength, and PEG grafting density. They observed that spheres exhibited the fastest internalization rate, followed by cubes, while rods and disks were the slowest. Many liposomal formulations have received clinical approval, like Doxil, Abraxane, etc. However, nanoparticles grafted with PEG prolong the systemic circulation of the particles and induces the EPR effect in tumor cells, but lack of target specificity often results in reduced therapeutic efficacy [27]. Because of that, more than 95% of passively targeted formulations fail to go bench to bedside.

3.2 Nanoparticle-based drug delivery by active targeting

An ideal nanoparticle delivery system should be proficient at reaching, recognizing, and delivering its payload to determined morbid tissues and avoid drug-induced toxicity to healthy tissues [7]. Therefore, functionalizing specific targeting moieties on the surface of nanoparticles is the most usual plan. Nanoparticles are functionalized on their outer surface by targeting moieties such as small molecule ligands, monoclonal antibodies, aptamers, cell-penetrating peptides, and proteins that are internalized into morbid cells by interacting with cell surface receptors like folate receptors, transferrin receptors, tyrosine kinases like EGFR, and so on [28] (**Figure 1**). Cell surface receptors that are significantly overexpressed in diseased cells, compared to normal healthy cells, provide a potential target for the design and development of actively targeted drug delivery and help to reduce off-target effects [7, 17]. These ligand moieties can interact with target-specific diseased cells and protect nanoparticles from enzymatic demolition.

Targeted drug delivery significantly minimizes the toxicity and induces patient compliance with less frequent dosing. Active targeting depends on ligands bound to the NP surface to improve their uptake selectivity and protect NPS from enzymatic destruction. The main principle of active targeting involves functionalizing an NP with a ligand that binds to a molecule overexpressed on cells. Ligands with a high binding affinity to a specific cell type exhibit higher delivery efficiency. One important thing to consider is that healthy cells still express the same molecule, and as healthy cells greatly outnumber, the chances of NPs missing their target will also increase. An intelligent selection and functionalization with multiple ligands can effectively mitigate the problem. Apart from this, active targeting mainly determined the kind of nanoparticle carrier, ligand targeting specific receptors, functional agents used for linking a ligand to the nanoparticles, hydrophilic polymers, and encapsulated active ingredients [28, 29].

Targeting tumor cell surface receptors is a common approach in active targeting. Nanoparticles were linked with targeted ligands for targeting specific cell receptors and thus upregulated the intracellular localization and therapeutic efficiency. Liposomes are conjugated with antibodies, a Y-shaped glycoprotein, or its fragments often termed as immunoliposomes, increasing the specificity of liposomes by targeting antigen-presenting cancer cells, which undergo endocytosis and destroy cancer cells followed by immune system clearance [28]. Folate receptors are membrane proteins overexpressed by various tumor cells. Folic acid is a ligand for targeting folate receptors, which pose high affinity, stability, and conjugation capacity [30]. It is conjugated with nanoparticles and a PEG spacer that inhibits steric hindrance between the cells and liposomes, which helps to increase cellular uptake and drug delivery of folate-targeted anticancer drugs. Targeting folate receptors with folic acid ligands helps deliver therapeutic and imaging agents effectively to the requisite site. Endothelial growth factor receptors (EGFR) overexpressed in solid tumors like non-small cell lung cancer, colorectal, squamous cell carcinoma of the ovary, kidney, head, neck, pancreas, prostate, and breast cancers can help in designing EGFR targeted drug delivery system. Antibody fragments used for targeting EGFR are functionalized on nanoparticle surfaces in order to acquire high targeting specificity [31]. Fibroblast growth factor receptors are overexpressed in cancers like lung, prostate, bladder, etc. Several groups have reported remarkable interaction of FGFs conjugated liposome with FGFR and discussed in detail [32, 33]. Overexpression of CD44 is observed in cancers like leukemia, ovarian, colon, gastric, pancreatic, and epithelial cancers. Hyaluronic acid acts as a ligand for CD44 and is used to deliver gemcitabine and DOX encapsulated within the liposomes [34].

Targeting the tumor microenvironment is another approach in active targeting, and one aspect is targeting the tumor vasculature instead of the tumor. This approach helps in the targeted destruction of neo-angiogenic blood vessels essential for tumor growth and metastasis [29, 35]. Vascular endothelial growth factor receptors (VEGFR) play a significant role in tumor angiogenesis and vascular permeability and regulate other aspects of tumorigenesis. Bevacizumab, a monoclonal antibody approved by USFDA, is used as an anti-human VEGF for targeting VEGFRs and FGFRs tyrosine receptors for active targeting [29]. Vascular cell adhesion molecules (VCAM-1) are cell adhesion molecules (CAMs) present on the endothelial cells responsible for inflammation. VCAM-1 is overexpressed in cancers like non-small cell lung cancer and tumor vasculature. Anti-VCAM and Fab-conjugated liposomes have high cellular uptake into Human Umbilical Vein and Endothelial Cells (HUVEC) compared to conventional liposomes [36].

Matrix metalloproteases (MMPs) are calcium-dependent endopeptidases involved in remodeling extracellular matrix, tumor invasiveness, and metastasis by modulating the formation of new blood vessels [37]. Conjugating MMP-2 cleavable peptides to liposomes loaded with cell-penetrating peptides increase the tumor selectivity. αβ-integrins are the heterodimeric transmembrane glycoproteins that facilitate the adhesion of endothelial cells with adjacent tissue and blood vessels. A tripeptide Arg-Gly-Asp (RGD) exhibited high specificity for $\alpha v\beta 3$ integrin helps in developing integrin targeted liposomes, which inhibits adhesion and angiogenesis in the tumor microenvironment (TME) [38]. Active targeting amends the intuitive patterns of a nanocarrier, directing to the specificity of the pathological tissue. In contrast, passive targeting delivery depends on the natural distribution of the therapeutic motifs and the EPR effect. Both the targeting mechanisms depend on blood circulation and the location of initial drug delivery. However, rare commercial advances are made using actively targeted NPs [39].

4. Nanomaterial and their application from biological analysis

4.1 Nanomaterial-driven faster and more accurate cell analysis

Early detection and diagnosis can play a pivotal role in the battle against many diseases. Scientists harness the unique attributes of nanomaterials to generate novel molecular contrast agents for *in vivo* imaging, sensing, measuring response to therapy, and liquid biopsy to study disease initiation, progression, and therapeutic response. Nanotechnology has a spacious range of accurate cell analyses. As described above, nanotechnology facilitates the development of desired formulations for individual cell analysis and their specific treatment applications, developing only one of its kind of applications for cell sensing/sensors, imaging, delivery, and diagnosis [39]. Since the importance of accurate cell analysis for nanoparticles is the latest approach, there is a big void for more discoveries and optimizations in various bio-applications.

4.2 Nanomaterial and *in vivo* imaging

The main lacunae in cancer treatment are a late diagnosis. The resolution of current imaging methods is low and can detect cancers at the late/ advanced stage or metastasized. A tissue biopsy can only help physicians to ascertain the tumor type and characteristics. Detection becomes even more challenging when metastatic modules and micrometastasis need to be identified. *In vivo* imaging enables us to non or minimally-invasively delve deep into the patient's tissue and is becoming increasingly popular for basic research and clinical applications. *In vivo,* molecular imaging focuses on obtaining spatiotemporal information about molecules of medical interest or biomarkers within a living body in real-time. Molecular *in vivo* imaging relies on contrast agents or medium that increases the contrast of physiological structure and enhances the sensitivity of detection. Different contrast agents are used for different *in vivo* imaging techniques including, radiocontrast, magnetic resonance imaging (MRI) contrast, ultrasound contrast, and optical contrast agents [40]. Precision diagnostics is dependent on high-resolution and high-contrast images. Nanomaterials are critical players in the generation of advanced contrast agents or media. Imageable nanoparticles can be classified based on their applications in nuclear, magnetic, optical, and acoustic imaging modalities. Moreover, NP-based contrast agents may be designed to integrate multiple detection modules and target specific cells. The advantages of nanoparticle-based contrast agents include enhanced specificity, increased photo and chemical stability, longer circulation time, engineered clearance pathways, and multimodal applications. The main *in vivo* imaging modalities include MRI, computed tomography (CT), positron emission tomography (PET), single-photon emission computed tomography (SPECT), ultrasonography (US), near-infrared fluorescence (NIRF), and two-photon intravital microscopy [41–43].

4.3 Nanoparticles as bio-sensors

By virtue of their unique properties, NPs make them ideal for their use for nano bio-sensing applications with enhanced sensitivity. Nanoparticles are widely used for detecting cells and pathogens, separating pathogens, recognize different biological substances, and detecting molecular and cellular functions [41, 42]. Accurate and professional separation of desired cells from the composite of various cell mixtures is essential for numerous biological applications. Nanoparticles have been investigated as a promising and very sensitive tool for the specific identification of cells. Identification and incarceration of metastatic cancer cells in the circulation can help understand and a strong analytical biomarker for various metastatic cancers, which can change the patient's prognosis. Nanoparticle-based methods are more frequently used for the identification and capture of metastatic circulating cancer cells. In this technique, magnetic nanoparticles were used to specifically track and separate the cells by using a ligand-receptor-based mechanism [42]. These techniques can also be used for the white blood cells with an anti-CD45-APC as a nanoparticle targeting ligand [44].

Additionally, various nanoparticle-based technologies have been investigated as a sensor for the identification and selection of various pathogens. The most frequently used method for finding bacteria is magnetic biosensors that involve immunological mechanisms using magnetic nanoparticles functionalized with antibodies against surface antigens. Many researchers have been utilizing small molecule tethered nanoparticles to analyze the bacteria successfully. Magnetic glyco-nanoparticles mediated particles could detect bacteria within 5 minutes, including subtraction from the sample by the bacterial interaction with carbohydrates on mammalian cell surfaces [41].

4.4 Nanoparticles as imaging agents

Nanoparticles have been investigated as imaging agents due to their exceptional physicochemical attributes for various biomedical applications such as cancers and cardiovascular diseases. Fluorescent labels can be easily conjugated to the surfaces of the nanoparticles by various chemical methods to design a wide range of imaging agents for dynamic *in vitro* and *in vivo* cellular imaging [45, 46]. Due to their passive and active targeting nature, nanoparticles can easily identify their specific biomarkers and accumulate at high concentrations in the targeted tissue. The high capability for nanoparticle modification and retention properties in the specific tissue region empowers their utilization as imaging amplifiers. Quantum dots are the most promising fluorescent labels for cellular imaging among all nanoparticles due to their inherent near infra region light emitting nature, reducing autofluorescence [47].

RGD peptide conjugated self-emitting quantum dots can be used for specific integrins highly expressed in tumors. The targeted nanoparticle has been examined for complex imaging competence, like imaging various molecular targets using different spectral emissions specific nanoparticles. Recently, nanotechnology has been used for imaging metastatic tumor cells in circulation, tumor cells, and their vasculature, stem cells, and lymph nodes [48]. Che et al. designed shortwave infrared window (SWIR)-responsive QDs for bone-specific real-time *in vivo* and *ex vivo* imaging and could visualize the significant bone structures Balb/C nude and Balb/C mouse [49]. The use of specific nanoparticles can help accurately decipher and image the gram-negative and gram-positive bacteria. Due to their fluorescence characteristics and specific bacterial cell wall interactions, they can be used in a wash-free fashion in bacterial imaging, which is significant for health care, food processing, and medical hygiene.

4.5 Application of nanoparticles in theranostics

Theranostic NMs are designed by the consolidation of diagnostic and therapeutic abilities in one biodegradable nanoparticle [50]. Novel theranostic materials should have the following properties; i) highly compatible with biological entities, ii) proficiently and precisely accumulate in desired morbid tissue, iii) describe the biochemical and morphological attributes of maladies, iv) exhibit minimal toxicological effects, v) and deliver a sufficient amount of therapeutic agent. Several techniques have been used to functionalize the surface of nanoparticles for theranostics use. Surface functionalization may include imaging agents, drugs, therapeutic cargo, nucleic acid, and contrast agents by either chemical functionalization or by biofunctionalization. Chemical functionalization depends on chemical cross-linking, while biofunctionalization of nanoparticles relies on bioinspired ligands obtained from natural phytochemicals). The use of nanotechnology offers a promising alternative for the diagnosis of various cancers. Various investigations convey that nanoparticles could be engineered for advanced diagnostic agents to detect cancers [51]. Double drug encapsulated liposomes can be functionalized to enhance theranostic efficacy [51, 52]. Multifunctional Metal nanoparticles can serve as a unique platform for cancer theranostics. The range of use of metal nanoparticles includes MRI imaging, biological catalysis, magnetic hyperthermia, magnetic drug delivery, photo-responsive drug delivery, and cell separation. Metal nanoparticles, including, Polymer-NP constructs containing Gd3+ complexes, Fe3 + – terpyridine complexes, and polymeric shell-based contrast agents, are widely studied for their theranostic use as MRI contrast. Magnetic particle imaging (MPI), a novel imaging technique, is based on the analysis of iron oxide NPs in response to a magnetic field.

Cheng et al. used GE11, a novel peptide with EGFR binding affinity and complexed with doxorubicin-loaded liposomes, and observed higher liposomal uptake and accumulation than, unconjugated liposomes using NIRF [53]. In another study Song et al. designed a multifunctional targeting liposome for targeting lung cancer. Octreotide (OCT), a synthetic 8-peptide analog of somatostatin, was used to surface coat the liposome for enhanced binding with the somatostatin receptors overexpressed in a subset of tumors. Double anti-cancer drug (Honokiol and epirubicin) co-encapsulated liposomes showed enhanced OCT- somatostatin receptor binding and *in vivo* response [54]. Cittadino et al. designed a theranostic long-circulating liposome with co-loaded prednisolone phosphate and an amphiphilic paramagnetic gadolinium contrast agent [Gd-DOTAMA(C18)(2)] for MRI monitoring of melanoma. The theranostically engineered liposomes showed long-term MRI-based detection without a loss in drug action [51]. The theranostic nanoparticle could assist in the patient's pre-selection, a prediction for responding to nanomedicine therapy. Moreover, nanomedicine-treated patients could be monitored throughout treatment duration while using nanomedicine formulations [39].

5. Biosafety and bioethics issues in handling mammalian cells

Biosafety is a notion that requires protecting human health and the surroundings of pathogenic and genetically modified mammalian cells or organisms used in the research. Mammalian cell culture is identified as a shelter for infectious etiologic substances, and it should change the compliance with containment measures recommended for the etiologic agent itself. The utility of cell cultures comes under the preview of a range of regulatory provisions that consider the estimation of biological risks. Genetically modified mammalian cell cultures were used in different continents; in that case, a bio-safety assessment should be regulated. The major guidelines issued to mitigate the biological risks for the users and environment are mainly by the World Health Organization; the Centers for Disease Control and Prevention, and the Swiss Expert Committee for Biosafety. Several countries or geographical zones have different directives; for example, in Europe, genetically modified research was brought into the regulatory provision (Directive 2009/41/EC). Mammalian cell culturing activities focusing on developing pharmaceutical drugs are covered by the Regulation (EC) No 726/2004 and its amendment laying down actions for the authorization and direction of medicinal goods for human and animal use. 3D cultures, especially organoid culture systems, are regularly used for disease modeling and studying nanomaterial-based physiological effects. Human Pluripotent stem cell-derived organoids are being generated from various human cell types and need better bio-safety and bioethics assessment. It must be ascertained that rules focusing on extenuating the biological risks for laboratory researchers, public health, and the environment falls under the preview one or several regulatory provisions based on biological risk assessment. Here, we are going to address the bio-safety issues involving mammalian cell cultures.

5.1 Bio-safety assessments of mammalian cell cultures

Biosafety refers to the way of protecting scientists, the health of other humans, and the environment from the probable side effects of microorganism, pathogenic, and genetically modified organisms and cells from human and mouse backgrounds. Laboratory biosafety uses safety principles and techniques to minimize the health hazard from accidental exposure or unplanned spillage while using infectious agents, toxins and other biological hazards in the laboratory setting. The bio-safety assessments applied to mammalian cells depend on a systematic assessment of the intrinsic attributes of the mammalian cultures like genetically modified cells and contaminated or intentionally infected with pathogens. **Figure 2** shows a summary of the biosafety assessment and management process that is followed while handling cell culture-based experiments. This also considers an exposure analysis, which means that type of exploitation carried out with the cultures should be considered. The risk analysis of cell cultures that carry the pathogens follows the same methods for analyzing pathogens themselves. Primarily, the inclusive depiction of major pathogens is measured by the subsequent guidelines (i) pathogenicity and the infectious dose (ii) mode of transmission, (iii) host range, (iv) the epidemiology, potential reservoir and vectors, and the ability to zoonosis (v) the stability and the resilience of the pathogens in the surroundings.

Moreover, information related to the physicochemical properties of the pathogenic organism is considered, such as (i) susceptibility to disinfectants, (ii) physical

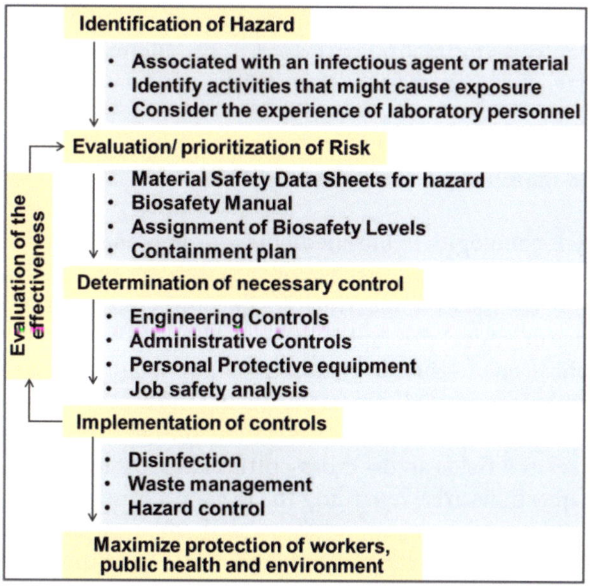

Figure 2.
Flow diagram illustrating the summarizing the biosafety assessment and management process while handling cell culture-based experiments. Flow chart is inspired by reference [55].

inactivation, and (iii) drug susceptibility (e.g., sensitivity and known resistance to antibiotics or antiviral compounds). Lastly, aspects related to the disease caused by the pathogen are also to be taken into consideration. This includes (i) the availability of effective prophylaxis, (ii) the availability of efficient therapy, and (iii) any reported case of laboratory-acquired infections (LAIs). Even though underemphasized, several LAIs of mammalian cell cultures (or having virus suspension) has appeared. Among all, the exposure to vaccinia viruses amplified in mammalian cell cultures causes infections to laboratory researchers. Guidelines have been developed recently to work cautiously with vaccinia viruses and take a count of LAIs relating to this virus [55].

Understanding and having a complete analysis of the intrinsic infections of cell cultures help to perform well and safe mammalian cell culture. To assess biological risks connected with the mammalian cell cultures, three intrinsic properties related to cell cultures should be considered: the species of origin, the cell type or type of tissue (the organ of origin of the cell line), and the status of the culture. Correspondingly, mammalian cells other than human cells render less risk; still, some infectious agents are proficient at crossing one species to another species, leading to zoonosis. Highly reported infections of viruses comprise hantavirus, hemorrhagic fever viruses, bird Influenza virus, and severe acute respiratory syndrome (SARS) associated virus. Primary cell cultures are created from organ tissues. Highly characterized mammalian cells give the lowest risks compared to primary cultures or less characterized cell lines. Mammalian cells originating from different laboratories without having any proof of identity may cause cross-contamination and pathogen spreading problems, and thereby proper risk assessment and cell characterization are warranted [55, 56]. Several techniques are available for the bio-safety assessment, like RT-PCR, flow cytometry, cytogenetic analysis, DNA fingerprinting, and iso-enzyme analysis. Adventitious contagions of mammalian cell cultures are a vital problem for any activity that involves cell culturing. Contamination agents for cell cultures are bacteria, fungi, mycoplasms, parasites, viruses, prions, and even other animal cells. Modulated experimental results suggest that they spoil the cell cultures. Bio-safety point of view modified mammalian cell cultures for laboratory research, production purposes, or diagnosis purposes they may give support for contaminating materials that cause harm to human health.

5.2 Bioethics and mammalian cell culture

The futuristic technologies in bio-medicine are changing the current concepts and opening up new dimensions. Interestingly as new optimistic channels are opening and expanding, the issues of bioethics are becoming accurate and pertinent. Bioethics is the use of ethical principles in the field of medicine and healthcare. The rational application of ethics in evaluating mammalian cell culture-based experiments is highly warranted, especially during the emerging waves of change in biomedicine. Increased International cross-connection to facilitate open discussion in bioethics and related fields across cross-cultural aspects in bioethics is vital [57]. Several relevant questions arise regarding the private and sensitive use of source data for cells, moral concerns regarding the uses of embryonic and fetal tissue, genetic manipulation, gene therapy, mixing of animal and human cells, tissue banking, legal and intellectual properties associated with *ex vivo* tissue-engineered cell-based products, an extension of human-ness, etc.

Regarding the humane use of animals, the National Institutes of Health has issued policies as mentioned in the Public Health Service Policy on Humane Care and Use of Laboratory Animals. FDA Human Tissue Task Force and the Center for Biologics Evaluation and Research (CBER) regulates the use of human cells or tissue for implantation, transplantation, infusion, or transfer into a human recipient. The International Society for Stem Cell Research (ISSCR) has also released guidelines for stem cell research and clinical translation. The United States Congress and state legislatures are instrumental in creating laws concerning bioethics. Several professional bioethics organizations, including the American Society for Bioethics and Humanities, American Society for Law, Medicine, and Ethics, Canadian Bioethics Society, provide a platform for discussion over bioethics [57]. Several public institutions supported by academicians and researcher-based initiative for propagating public dialog plays a vital role in educating the masses.

6. Significance of IPR on industrial and academic scale

Intellectual property rights (IPR) prevail in any primitive design of the human brain, such as methodical design. IPR mentions the lawful rights agreed to the designer for guarding his innovation for a definite period. These lawful rights grant special rights to the originator or his lender to exploit his idea for a specific period. It is well established that IPR participates in the financial system. It is furthermore overwhelmingly recognized that the intellectualism linked with the originality must be agreed due to value so that products come out of intelligence. The importance of the producer of the technology has turn into lofty and consequently guard the information against unauthorized persons, the use has become a measure, at least sometimes, that would make sure revitalization of the research, investments in developing the technology. IPR helps to look after funds, time, capital, endeavor invested by the producer of an intellectual idea; as a result, IPR, in this way, encourages the profitable encouragement of a realm by encouraging positive competition and heartening trade and industry [58].

The industries have reputations in discussions about IPR strategies, and they are in the face line for controversies about the association among IPRs, R&D incentives, cost, and right to use to supplies [59]. Although, some discussions on the critical issue are relatively little practical proof to support developing IPR policy. This experimental evidence on IP and products inspect practical issues are the primary sources of the data. The industrial sector is composite and much synchronized in the majority of economies. Looking cross-nationally, the contrast among the countries in their perspective on these essential policy affairs generates some additional provocations. In a cosmopolitan industry having control over research and development conveniences in many countries, anticipating a successful transnational technology, goods are raised and developed internationally and are commercialized worldwide. Still, retails are nationalized, with no considerable uniformity across the nations in IPR authorities and various public health care organizations. IPRs may shore up significant discrepancies to price across the nations in returns and demand to prices. These discrepancies in the prices may potentially develop new local and global disagreements. Prominently, for any nation, the essential exchange in IPR regulation options is incredibly dependent on the organizations and function of its health care system.

While having a commendable collaboration, the complete fulfillment of a patent portfolio is to give equal rights for industry and academic institutions. In many countries, research organizations pursuing research in academic institutions, despite their most important work in society as a generator of the intellectual idea, the main concern is to be to deal with IP in a proficient mode. All academic institutions must become accustomed to this development to successfully fulfill the responsibility entrusted to a national or regional innovation ecosystem.

On the supply side, goods safety, supervision of manufacturing, and legal frameworks leading technology transfer among public-funded academic institutions and money-making industries playing an equal role in determining competition. Providing IPR policy to academic institutions has a favorable outcome and various settlements for shareholders. The most significant overarching advantage of these IPR policies was pronounced increases involvement in improving the global innovation performance, i.e., ultimately leading to improving the marketable products and processes. The development of spin-out companies from universities is also growing at a faster rate. The critical part is that the university should own the background IP. Then a resource of external financial support is necessary to finance the start-up company. IPR affairs at academic institutions glow enormous meandering return impending for the national economy. Publishing articles regarding innovations play an essential role in the profession of academic scientists. Participating in knowledge transfer from academia to manufacturing industries can promote academic entrepreneurship. Moreover, these patents have precious information than other publishing articles. Thus, utilizing and increasing patent writing might be beneficial in scientific research. Appropriate IPR policies and tractable technology transfer professionals play a pivotal role in streamlining the necessary work-frame. Published patents improve the economy and reputation of the academic institutions as well as the researchers.

7. Conclusion

Nanomaterials, due to their nano-size and unique physicochemical properties, have contributed significantly to the advance of biomedicine. The scope of nanomedicine also relies on the intelligent engineering of different nanoparticles with tunable attributes to modulate their nano-bio communications for biomedical applications. Elucidation of nanoparticle interactions with biological systems will help find favorable physicochemical properties to enhance biocompatibility and therapeutic efficacy with no adverse effects. A complete toxicological evaluation of engineered nanomaterials is still inadequately understood, restraining the successful translation of nanomedicine. Nanoparticle surface functionalization with specific targeting moieties can effectively develop ideal nanoparticle delivery systems for various biomedical applications and targeted therapeutics. Hence, *in vitro* 2D and 3D cell culture systems can accelerate biocompatibility and biotoxicity studies to drive the disease-specific application of nanoparticles [60]. Nanoparticles are progressively used in a wide variety of cell and tissue-specific biological analyses, including cell analysis, in vivo imaging, biosensors, and theranostics. Hence the issue of biosafety and bioethics has become a vital issue while using mammalian cell cultures. This chapter summarizes the critical aspects of biosafety and bioethics associated with nanomaterial-associated studies.

In conclusion, MCC is an essential tool in modern-day biomedicine, and its applications are countless in the diagnosis and therapy of human diseases. Cell culture procedures are reliable, reproducible, and unbiased, but culturing the cells is complex at times. The vast opportunities to employ MCC procedures to address rudimentary and translational research queries have elucidated the essential attentions for setting up a cell culture laboratory. Especially 3D organoid culture methods have created a cellular environment that mimics the *in vivo* environment.

Genome sequencing, mapping, and annotating its genetic code have become a priority in biotechnology, especially intending to understand the interaction of nanoparticles and mammalian cells. Reporting and cataloging the identified gene sequences can be critical for the progress of science and also for disease-specific therapeutics. Nanotechnology-based research has contributed significantly to many scientific fields and associated industries. Hence nanotechnology, combined with the mammalian cell culture system, can result in a research solution and can deliver considerable benefits to society at large. Hence the importance of intellectual property rights for protecting the innovator's right over the discovery. A good understanding of the IPR policies and technology transfer protocol is vital. Academic institutions and government organizations can assist in creating a congenial platform for efficient policy management. A deeper understanding of nanoparticle-cell interaction and the design of futuristic nanocarriers can open up an era of next-generation therapeutics and theranostics.

Acknowledgements

M.K.P. acknowledges Professors S. Dubinett, B. Gomperts, and V. Hartenstein from UCLA for providing constant support and mentoring.

Author contributions

Conceptualization - MKP; writing original draft preparation - HR, AM, MKP; Review and editing - MKP.

Funding

No funds available.

Conflicts of interest

The authors declare no conflict of interest. The authors have no other pertinent affiliations or financial connection with any organization or entity with a financial interest in or financial conflict with the subject matter or materials discussed in the manuscript apart from those disclosed.

Abbreviations

CBER	Centre for Biologics Evaluation and Research
CT	Computed Tomography
EGFR	Epidermal Growth Factor Receptor
EPR	Enhanced Permeability and Retention
FGFR	Fibroblast Growth Factor Receptor
HUVEC	Human Umbilical Vein Endothelial Cells
IgM	Immunoglobulin M
IPR	Intellectual Property Rights
ISSCR	International Society for Stem Cell Research
LAIs	Laboratory-Acquired Infections
MMPs	Matrix Metalloproteases
MPI	Magnetic Particle imaging
MRI	Magnetic Resonance Imaging
NIH	National Institutes of Health
NMs	Nanomaterials
NPs	Nanoparticles
NRIF	Near-infrared Fluorescence
NT	Nanotechnology
OCT	Octreotide
PEG	Polyethylene glycol
PET	Positron Emission Tomography
QD	Quantum Dots
SARS	Severe Acute Respiratory Syndrome
SPECT	Single-photon Emission Computed Tomography
SRA	Sequence Read Archive
SWIR	Shortwave Infrared
TME	Tumor Microenvironment
US	Ultrasonography
USFDA	United States Food and Drug Administration
VCAM-1	Vascular Cell Adhesion Molecules-1
VEGFR	Vascular Endothelial Growth Factor Receptors

Author details

Harikrishnareddy Rachamalla[1], Anubhab Mukherjee[2*] and Manash K. Paul[3*]

1 Mayo Clinic Jacksonville, Florida, USA

2 Esperer Onco Nutrition Pvt. Ltd., Hyderabad, India

3 Division of Pulmonary and Critical Care Medicine, David Geffen School of Medicine, University of California, Los Angeles (UCLA), Los Angeles, CA, USA

*Address all correspondence to: dranubhab@esperernutrition.com and manashp@ucla.edu

References

[1] Poon W, Kingston BR, Ouyang B, Ngo W, Chan WCW. A framework for designing delivery systems. *Nat Nanotechnol* 2020;**15**(10):819-829. [10.1038/s41565-020-0759-5]

[2] Golombek SK, May JN, Theek B, et al. Tumor targeting via EPR: Strategies to enhance patient responses. *Adv Drug Deliv Rev* 2018;**130**:17-38.[10.1016/j.addr.2018.07.007]

[3] Mukherjee S, Madamsetty VS, Bhattacharya D, Roy Chowdhury S, Paul MK, Mukherjee A. Recent Advancements of Nanomedicine in Neurodegenerative Disorders Theranostics. *Advanced Functional Materials* 2020;**30**(35).[10.1002/adfm.202003054]

[4] Freshney RI. *Culture of Animal Cells.* [10.1002/9780470649367]

[5] Stacey G. Current Developments in Cell Culture Technology. In: *New Technologies for Toxicity Testing*. 2012;1-13. [10.1007/978-1-4614-3055-1_1]

[6] Wilkinson DC, Alva-Ornelas JA, Sucre JMS, et al. Development of a Three-Dimensional Bioengineering Technology to Generate Lung Tissue for Personalized Disease Modeling. *STEM CELLS Translational Medicine* 2017;**6**(2):622-633.[10.5966/sctm.2016-0192]

[7] Sanna V, Sechi M. Therapeutic Potential of Targeted Nanoparticles and Perspective on Nanotherapies. *ACS Medicinal Chemistry Letters* 2020;**11**(6):1069-1073.[10.1021/acsmedchemlett.0c00075]

[8] Mitchell MJ, Billingsley MM, Haley RM, Wechsler ME, Peppas NA, Langer R. Engineering precision nanoparticles for drug delivery. *Nature Reviews Drug Discovery* 2020;**20**(2):101-124.[10.1038/s41573-020-0090-8]

[9] Li X, Wang L, Fan Y, Feng Q, Cui F-z. Biocompatibility and Toxicity of Nanoparticles and Nanotubes. *Journal of Nanomaterials* 2012;**2012**:1-19. [10.1155/2012/548389]

[10] Patra JK, Das G, Fraceto LF, et al. Nano based drug delivery systems: recent developments and future prospects. *Journal of Nanobiotechnology* 2018;**16**(1).[10.1186/s12951-018-0392-8]

[11] Wagner AM, Knipe JM, Orive G, Peppas NA. Quantum dots in biomedical applications. *Acta Biomaterialia* 2019;**94**:44-63.[10.1016/j.actbio.2019.05.022]

[12] Beltrán-Gracia E, López-Camacho A, Higuera-Ciapara I, Velázquez-Fernández JB, Vallejo-Cardona AA. Nanomedicine review: clinical developments in liposomal applications. *Cancer Nanotechnology* 2019;**10**(1).[10.1186/s12645-019-0055-y]

[13] Bulbake U, Doppalapudi S, Kommineni N, Khan W. Liposomal Formulations in Clinical Use: An Updated Review. *Pharmaceutics* 2017;**9**(4).[10.3390/pharmaceutics9020012]

[14] Liu C, Zhang L, Zhu W, et al. Barriers and Strategies of Cationic Liposomes for Cancer Gene Therapy. *Molecular Therapy - Methods & Clinical Development* 2020;**18**:751-764.[10.1016/j.omtm.2020.07.015]

[15] Jeevanandam J, Barhoum A, Chan YS, Dufresne A, Danquah MK. Review on nanoparticles and nanostructured materials: history, sources, toxicity and regulations. *Beilstein Journal of Nanotechnology* 2018;**9**:1050-1074.[10.3762/bjnano.9.98]

[16] Zein R, Sharrouf W, Selting K. Physical Properties of Nanoparticles

That Result in Improved Cancer Targeting. *Journal of Oncology* 2020;**2020**:1-16.[10.1155/2020/5194780]

[17] Blanco E, Shen H, Ferrari M. Principles of nanoparticle design for overcoming biological barriers to drug delivery. *Nature Biotechnology* 2015;**33**(9):941-951.[10.1038/nbt.3330]

[18] Huang Y-W, Cambre M, Lee H-J. The Toxicity of Nanoparticles Depends on Multiple Molecular and Physicochemical Mechanisms. *International Journal of Molecular Sciences* 2017;**18**(12).[10.3390/ijms18122702]

[19] Sanità G, Carrese B, Lamberti A. Nanoparticle Surface Functionalization: How to Improve Biocompatibility and Cellular Internalization. *Frontiers in Molecular Biosciences* 2020;**7**.[10.3389/fmolb.2020.587012]

[20] Gonda A, Zhao N, Shah JV, et al. Engineering Tumor-Targeting Nanoparticles as Vehicles for Precision Nanomedicine. *Med One* 2019;**4**.[10.20900/mo.20190021]

[21] Rawat M, Yadukrishnan P, Kumar N. Mechanisms of Action of Nanoparticles in Living Systems. In: *Microbial Biotechnology in Environmental Monitoring and Cleanup*. 2018;220-236.[10.4018/978-1-5225-3126-5.ch014]

[22] Abbas Z, Rehman S. An Overview of Cancer Treatment Modalities. In: *Neoplasm*. 2018.[10.5772/intechopen.76558]

[23] Chenthamara D, Subramaniam S, Ramakrishnan SG, et al. Therapeutic efficacy of nanoparticles and routes of administration. *Biomaterials Research* 2019;**23**(1).[10.1186/s40824-019-0166-x]

[24] Kim ST, Saha K, Kim C, Rotello VM. The Role of Surface Functionality in Determining Nanoparticle Cytotoxicity. *Accounts of Chemical Research* 2013;**46**(3):681-691.[10.1021/ar3000647]

[25] Albanese A, Tang PS, Chan WCW. The Effect of Nanoparticle Size, Shape, and Surface Chemistry on Biological Systems. *Annual Review of Biomedical Engineering* 2012;**14**(1):1-16.[10.1146/annurev-bioeng-071811-150124]

[26] Salmaso S, Caliceti P. Stealth Properties to Improve Therapeutic Efficacy of Drug Nanocarriers. *Journal of Drug Delivery* 2013;**2013**:1-19.[10.1155/2013/374252]

[27] Gal N, Charwat V, Städler B, Reimhult E. Poly(ethylene glycol) Grafting of Nanoparticles Prevents Uptake by Cells and Transport Through Cell Barrier Layers Regardless of Shear Flow and Particle Size. *ACS Biomaterials Science & Engineering* 2019;**5**(9):4355-4365.[10.1021/acsbiomaterials.9b00611]

[28] Large DE, Soucy JR, Hebert J, Auguste DT. Advances in Receptor-Mediated, Tumor-Targeted Drug Delivery. *Advanced Therapeutics* 2019;**2**(1).[10.1002/adtp.201800091]

[29] Mukherjee A, Madamsetty VS, Paul MK, Mukherjee S. Recent Advancements of Nanomedicine towards Antiangiogenic Therapy in Cancer. *International Journal of Molecular Sciences* 2020;**21**(2).[10.3390/ijms21020455]

[30] Frigerio B, Bizzoni C, Jansen G, et al. Folate receptors and transporters: biological role and diagnostic/therapeutic targets in cancer and other diseases. *Journal of Experimental & Clinical Cancer Research* 2019;**38**(1).[10.1186/s13046-019-1123-1]

[31] Siwak DR, Carey M, Hennessy BT, et al. Targeting the Epidermal Growth Factor Receptor in Epithelial Ovarian Cancer: Current Knowledge and Future Challenges. *Journal of Oncology* 2010;**2010**:1-20.[10.1155/2010/568938]

[32] Terada T, Mizobata M, Kawakami S, Yamashita F, Hashida M. Optimization of tumor-selective targeting by basic fibroblast growth factor-binding peptide grafted PEGylated liposomes. *Journal of Controlled Release* 2007;**119**(3):262-270.[10.1016/j.jconrel.2007.01.018]

[33] Holzmann, Marian. Importance of Translational Research for Targeting Fibroblast Growth Factor Receptor Signaling in Cancer. *Cells* 2019;**8**(10).[10.3390/cells8101191]

[34] Mukherjee A, Waters AK, Babic I, et al. Antibody drug conjugates: Progress, pitfalls, and promises. *Human Antibodies* 2018;**27**(1):53-62.[10.3233/hab-180348]

[35] Roma-Rodrigues C, Mendes R, Baptista P, Fernandes A. Targeting Tumor Microenvironment for Cancer Therapy. *International Journal of Molecular Sciences* 2019;**20**(4).[10.3390/ijms20040840]

[36] Sakurai, Akita, Harashima. Targeting Tumor Endothelial Cells with Nanoparticles. *International Journal of Molecular Sciences* 2019;**20**(23).[10.3390/ijms20235819]

[37] Silva S, Almeida A, Vale N. Combination of Cell-Penetrating Peptides with Nanoparticles for Therapeutic Application: A Review. *Biomolecules* 2019;**9**(1).[10.3390/biom9010022]

[38] Kapp TG, Rechenmacher F, Neubauer S, et al. A Comprehensive Evaluation of the Activity and Selectivity Profile of Ligands for RGD-binding Integrins. *Scientific Reports* 2017;**7**(1).[10.1038/srep39805]

[39] Maghsoudnia N, Eftekhari RB, Sohi AN, Zamzami A, Dorkoosh FA. Application of nano-based systems for drug delivery and targeting: a review. *Journal of Nanoparticle Research* 2020;**22**(8).[10.1007/s11051-020-04959-8]

[40] Issa B, M. Obaidat I. Magnetic Nanoparticles as MRI Contrast Agents. In: *Magnetic Resonance Imaging.* 2019.[10.5772/intechopen.84649]

[41] Colino C, Millán C, Lanao J. Nanoparticles for Signaling in Biodiagnosis and Treatment of Infectious Diseases. *International Journal of Molecular Sciences* 2018;**19**(6).[10.3390/ijms19061627]

[42] Zhang R, Belwal T, Li L, Lin X, Xu Y, Luo Z. Nanomaterial-based biosensors for sensing key foodborne pathogens: Advances from recent decades. *Comprehensive Reviews in Food Science and Food Safety* 2020;**19**(4):1465-1487.[10.1111/1541-4337.12576]

[43] Lin Q, Fathi P, Chen X. Nanoparticle delivery in vivo: A fresh look from intravital imaging. *EBioMedicine* 2020;**59**.[10.1016/j.ebiom.2020.102958]

[44] Su Y, Xie Z, Kim GB, Dong C, Yang J. Design Strategies and Applications of Circulating Cell-Mediated Drug Delivery Systems. *ACS Biomaterials Science & Engineering* 2015;**1**(4):201-217.[10.1021/ab500179h]

[45] Wolfbeis OS. An overview of nanoparticles commonly used in fluorescent bioimaging. *Chemical Society Reviews* 2015;**44**(14):4743-4768.[10.1039/c4cs00392f]

[46] van der Meel R, Sulheim E, Shi Y, Kiessling F, Mulder WJM, Lammers T. Smart cancer nanomedicine. *Nature Nanotechnology* 2019;**14**(11):1007-1017.[10.1038/s41565-019-0567-y]

[47] Benito-Alifonso D, Richichi B, Baldoneschi V, et al. Quantum Dot-Based Probes for Labeling and Imaging of Cells that Express Matrix Metalloproteinases. *ACS Omega*

2018;**3**(8):9822-9826.[10.1021/acsomega.8b00633]

[48] Siemer S, Wünsch D, Khamis A, et al. Nano Meets Micro-Translational Nanotechnology in Medicine: Nano-Based Applications for Early Tumor Detection and Therapy. *Nanomaterials* 2020;**10**(2).[10.3390/nano10020383]

[49] Che Y, Feng S, Guo J, et al. In vivo live imaging of bone using shortwave infrared fluorescent quantum dots. *Nanoscale* 2020;**12**(43):22022-22029.[10.1039/d0nr06261h]

[50] Mukherjee A, Paul M, Mukherjee S. Recent Progress in the Theranostics Application of Nanomedicine in Lung Cancer. *Cancers* 2019;**11**(5).[10.3390/cancers11050597]

[51] Madamsetty VS, Paul MK, Mukherjee A, Mukherjee S. Functionalization of Nanomaterials and Their Application in Melanoma Cancer Theranostics. *ACS Biomaterials Science & Engineering* 2019;**6**(1):167-181.[10.1021/acsbiomaterials.9b01426]

[52] Agrawal V, Paul MK, Mukhopadhyay AK. 6-Mercaptopurine and Daunorubicin Double Drug Liposomes—Preparation, Drug-Drug Interaction and Characterization. *Journal of Liposome Research* 2008;**15**(3-4):141-155.[10.1080/08982100500364081]

[53] Chen D-W, Cheng L, Huang F, et al. GE11-modified liposomes for non-small cell lung cancer targeting: preparation, ex vitro and in vivo evaluation. *International Journal of Nanomedicine* 2014;10.2147/ijn.S53310.[10.2147/ijn.S53310]

[54] Song X-l, Ju R-j, Xiao Y, et al. Application of multifunctional targeting epirubicin liposomes in the treatment of non-small-cell lung cancer. *International Journal of Nanomedicine* 2017;**12**:7433-7451.[10.2147/ijn.S141787]

[55] Herman P, Pauwels K. Biosafety Recommendations on the Handling of Animal Cell Cultures. In: *Animal Cell Culture*. 2015:689-716.[10.1007/978-3-319-10320-4_22]

[56] Geraghty RJ, Capes-Davis A, Davis JM, et al. Guidelines for the use of cell lines in biomedical research. *British Journal of Cancer* 2014;**111**(6):1021-1046.[10.1038/bjc.2014.166]

[57] Segers S, Mertes H, de Wert G, Dondorp W, Pennings G. Balancing Ethical Pros and Cons of Stem Cell Derived Gametes. *Annals of Biomedical Engineering* 2017;**45**(7):1620-1632.[10.1007/s10439-017-1793-9]

[58] Prabu SL, Suriyaprakash TNK, Thirumurugan R. Introductory Chapter: Intellectual Property Rights. In: *Intellectual Property Rights.* 2017.[10.5772/intechopen.69359]

[59] Lemley MA. *Intellectual property in the new technological age : 2016.* Berkeley, CA: Clause 8 Pub.; 2016

[60] Abdel Fattah AR, Ranga A. Nanoparticles as Versatile Tools for Mechanotransduction in Tissues and Organoids. *Frontiers in Bioengineering and Biotechnology* 2020;**8**.[10.3389/fbioe.2020.00240]